MATTER AND MOTION

MAXWELL PRESS

MATTER AND MOTION

J Clerk Maxwell M.A., LL.D.EDIN., F.R.SS.L. & E.

Honorary Fellow of Trinity College, and Professor of Experimental Physics in the University of Cambridge

Sir Joseph Larmor, F.R.S., M.P.

Fellow of St John's College, and Lucasian Professor of Mathematics

MAXWELL PRESS

Chennai Trichy NewDelhi

MAXWELL PRESS

This edition has been published in india by arrangement with Carson Books, UK

ISBN 978-93-90063-57-4 Maxwell Press

All rights reserved No. 44, Nallathambi Street,
Triplicane, Chennai 600 005

MJP 903 © Publishers, 2023

Publisher's Note

The legacy of a country is in its varied cultural heritage, historical literature, developments in the field of economy and science. The top nations in the world are competing in the field of science, economy and literature. This vast legacy has to be conserved and documented so that it can be bestowed to the future generation. The knowledge of this legacy is slowly getting perished in the present generation due to lack of documentation.

Keeping this in mind, the concern with retrospective acquiring of rare books has been accented recently by the burgeoning reprint industry. Maxwell Press is gratified to retrieve the rare collections with a view to bring back those books that were landmarks in their time.

In this effort, a series of rare books would be republished under the banner, "Maxwell Press". The books in the reprint series have been carefully selected for their contemporary usefulness as well as their historical importance within the intellectual. We reconstruct the book with slight enhancements made for better presentation, without affecting the contents of the original edition.

Most of the works selected for republishing covers a huge range of subjects, from history to anthropology. We believe this reprint edition will be a service to the numerous researchers and practitioners active in this fascinating field. We allow readers to experience the wonder of peering into a scholarly work of the highest order and seminal significance.

MAXWELL PRESS

PREFACE (1877)

PHYSICAL SCIENCE, which up to the end of the eighteenth century had been fully occupied in forming a conception of natural phenomena as the result of forces acting between one body and another, has now fairly entered on the next stage of progress—that in which the energy of a material system is conceived as determined by the configuration and motion of that system, and in which the ideas of configuration, motion, and force are generalised to the utmost extent warranted by their physical definitions.

To become acquainted with these fundamental ideas, to examine them under all their aspects, and habitually to guide the current of thought along the channels of strict dynamical reasoning, must be the foundation of the training of the student of Physical Science.

The following statement of the fundamental doctrines of Matter and Motion is therefore to be regarded as an introduction to the study of Physical Science in general.

NOTE

In this reprint of Prof. Clerk Maxwell's classical tractate on the principles of dynamics, the changes have been confined strictly to typographical and a few verbal improvements. After trial, the conclusion has been reached that any additions to the text would alter the flavour of the work, which would then no longer be characteristic of its author. Accordingly only brief footnotes have been introduced: and the few original footnotes have been distinguished from them by Arabic numeral references instead of asterisks and other marks. A new index has been prepared.

A general exposition of this kind cannot be expected, and doubtless was not intended, to come into use as a working textbook: for that purpose methods of systematic calculation must be prominent. But as a reasoned conspectus of the Newtonian dynamics, generalizing gradually from simple particles of matter to physical systems which are beyond complete analysis, drawn up by one of the masters of the science, with many interesting side-lights, it must retain its power of suggestion even though parts of the vector exposition may now seem somewhat abstract. The few critical footnotes and references to Appendices that have been added may help to promote this feature of suggestion and stimulus.

The treatment of the fundamental principles of dynamics has however been enlarged on the author's own lines by the inclusion of the Chapter "On the Equations of Motion of a Connected System" from vol. ii of *Electricity and Magnetism*. For permission to make use of this chapter the thanks of the publishers are due to the Clarendon Press of the University of Oxford.

With the same end in view two Appendices have been added by the editor. One of them treats the Principle of Relativity of motion, which has recently become very prominent in wider physical connexions, on rather different lines from those in the text. The other aims at development of the wider aspects of the Principle of Least Action, which has been asserting its position more and more as the essential principle of connexion between the various domains of Theoretical Physics.

These additions are of course much more advanced than the rest of the book: but they will serve to complete it by presenting the analytical side of dynamical science, on which it justly aspires to be the definite foundation for all Natural Philosophy.

The editor desires to express his acknowledgment to the Cambridge University Press, and especially to Mr J. B. Peace, for assistance and attention.

J. L.

BIOGRAPHICAL NOTE

JAMES CLERK MAXWELL was born in Edinburgh in 1831, the only son of John Clerk Maxwell, of Glenlair, near Dalbeattie, a family property in south-west Scotland to which the son succeeded. After an early education at home, and at the University of Edinburgh, he proceeded to Cambridge in 1850, first to Peterhouse, migrating afterwards to Trinity College. In the Mathematical Tripos of 1854, the Senior Wrangler was E. J. Routh, afterwards a mathematical teacher and investigator of the highest distinction, and Clerk Maxwell was second: they were placed as equal soon after in the Smith's Prize Examination.

He was professor of Natural Philosophy at Aberdeen from 1856 to 1860, in King's College, London from 1860 to 1865, and then retired to Glenlair for six years, during which the teeming ideas of his mind doubtless matured and fell into more systematic forms. He was persuaded to return into residence at Cambridge in 1871, to undertake the task of organizing the new Cavendish Laboratory. But after a time his health broke, and he died in 1879 at the age of 48 years.

His scientific reputation during his lifetime was upheld mainly by British mathematical physicists, especially by the Cambridge school. But from the time that Helmholtz took up the study of his theory of electric action and light in 1870, and discussed it in numerous powerful memoirs, the attention given abroad to his work gradually increased, until as in England it became the dominating force in physical science.

Nowadays by universal consent his ideas, as the mathematical interpreter and continuator of Faraday, rank as the greatest advance in our understanding of the laws of the physical universe that has appeared since the time of Newton. As with Faraday, his profound investigations into nature were concomitant with deep religious reverence for nature's cause. See the *Life* by L. Campbell and W. Garnett (Macmillan, 1882). The treatise on *Electricity and Magnetism* and the *Theory of Heat* contain an important part of his work. His Scientific Papers were republished by the Cambridge University Press in two large memorial volumes. There are many important letters from him in the *Memoir and Scientific Correspondence of Sir George Stokes*, Cambridge, 1904.

The characteristic portrait here reproduced, perhaps for the first time, is from a *carte de visite* photograph taken probably during his London period.

J. L.

CONTENTS

CHAPTER I

INTRODUCTION

CHAPTER II

ON MOTION

CHAPTER III

ON FORCE

CHAPTER IV

ON THE PROPERTIES OF THE CENTRE OF MASS OF A MATERIAL SYSTEM

CHAPTER V

ON WORK AND ENERGY

CHAPTER VI

RECAPITULATION

CHAPTER VII

THE PENDULUM AND GRAVITY

CHAPTER VIII

UNIVERSAL GRAVITATION

MATTER AND MOTION

CHAPTER I

INTRODUCTION

1. Nature of Physical Science

Physical Science is that department of knowledge which relates to the order of nature, or, in other words, to the regular succession of events.

The name of physical science, however, is often applied in a more or less restricted manner to those branches of science in which the phenomena considered are of the simplest and most abstract kind, excluding the consideration of the more complex phenomena, such as those observed in living beings.

The simplest case of all is that in which an event or phenomenon can be described as a change in the arrangement of certain bodies. Thus the motion of the moon may be described by stating the changes in her position relative to the earth in the order in which they follow one another.

In other cases we may know that some change of arrangement has taken place, but we may not be able to ascertain what that change is.

Thus when water freezes we know that the molecules or smallest parts of the substance must be arranged differently in ice and in water. We also know that this arrangement in ice must have a certain kind of symmetry, because the ice is in the form of symmetrical crystals, but we have as yet no precise knowledge of the actual arrangement of the molecules in ice. But whenever we can completely describe the change of arrangement we have a knowledge, perfect so far as it extends, of what has taken place, though we may still have to learn the necessary conditions under which a similar event will always take place.

Hence the first part of physical science relates to the relative position and motion of bodies.

2. Definition of a Material System

In all scientific procedure we begin by marking out a certain region or subject as the field of our investigations. To this we must confine our attention, leaving the rest of the universe out of account till we have completed the investigation in which we are engaged. In physical science, therefore, the first step is to define clearly the material system which we make the subject of our statements. This system may be of any degree of complexity. It may be a single material particle, a body of finite size, or any number of such bodies, and it may even be extended so as to include the whole material universe.

3. Definition of Internal and External

All relations or actions between one part of this system and another are called Internal relations or actions.

Those between the whole or any part of the system and bodies not included in the system are called External relations or actions. These we study only so far as they affect the system itself, leaving their effect on external bodies out of consideration. Relations and actions between bodies not included in the system are to be left out of consideration. We cannot investigate them except by making our system include these other bodies.

4. Definition of Configuration

When a material system is considered with respect to the relative position of its parts, the assemblage of relative positions is called the Configuration of the system.

A knowledge of the configuration of the system at a given instant implies a knowledge of the position of every point of the system with respect to every other point at that instant.

5. Diagrams

The configuration of material systems may be represented in models, plans, or diagrams. The model or diagram is supposed to resemble the material system only in form, not necessarily in any other respect.

A plan or a map represents on paper in two dimensions what may really be in three dimensions, and can only be completely represented by a model. We shall use the term Diagram to signify any geometrical figure, whether plane or not, by means of which we study the properties of a material system. Thus, when we speak of the configuration of a system, the image which we form in our minds is that of a diagram, which completely represents the configuration, but which has none of the other properties of the material system. Besides diagrams of configuration we may have diagrams of velocity, of stress, etc., which do not represent the form of the system, but by means of which its relative velocities or its internal forces may be studied.

6. A Material Particle

A body so small that, *for the purposes of our investigation*, the distances between its different parts may be neglected, is called a material particle.

Thus in certain astronomical investigations the planets, and even the sun, may be regarded each as a material particle, because the difference of the actions of different parts of these bodies does not come under our notice. But we cannot treat them as material particles when we investigate their rotation. Even an atom, when we consider it as capable of rotation, must be regarded as consisting of many material particles.

The diagram of a material particle is of course a mathematical point, which has no configuration.

7. Relative Position of two Material Particles

The diagram of two material particles consists of two points, as, for instance, A and B.

The position of B relative to A is indicated by the direction and length of the straight line $\overline{AB}$ drawn *from A to B*. If you start from A and travel in the direction indicated by the line $\overline{AB}$ and for a distance equal to the length of that line, you will get to B. This direction and distance may be indicated equally well by any other line, such as $\overline{ab}$, which is parallel and equal to $\overline{AB}$. The position of A with respect to B is indicated by the direction and length of the line $\overline{BA}$, drawn *from B to A*, or the line $\overline{ba}$, equal and parallel to $\overline{BA}$.

It is evident that $\overline{BA} = -\overline{AB}$.

In naming a line by the letters at its extremities, the order of the letters is always that in which the line is to be drawn.

8. Vectors

The expression $\overline{AB}$, in geometry, is merely the name of a line. Here it indicates the operation by which the line is drawn, that of carrying a tracing point in a certain direction for a certain distance. As indicating an operation, AB is called a Vector, and the operation is completely defined by the direction and distance of the transference. The starting point, which is called the Origin of the vector, may be anywhere.

To define a finite straight line we must state its origin as well as its direction and length. All vectors, however, are regarded as equal which are parallel (and drawn towards the same parts) and of the same magnitude.

Any quantity, such, for instance, as a velocity or a

force*, which has a definite direction and a definite magnitude may be treated as a vector, and may be indicated in a diagram by a straight line whose direction is parallel to the vector, and whose length represents, according to a determinate scale, the magnitude of the vector.

9. System of Three Particles

Let us next consider a system of three particles.

Its configuration is represented by a diagram of three points, A, B, C.

The position of B with respect to A is indicated by the vector $\overline{AB}$, and that of C with respect to B by the vector $\overline{BC}$.

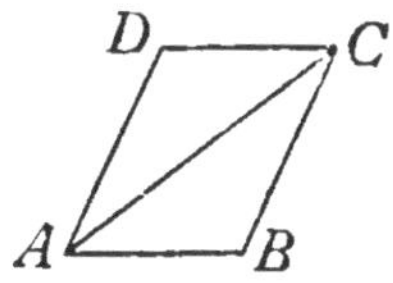

Fig. 1.

It is manifest that from these data, when A is known, we can find B and then C, so that the configuration of the three points is completely determined.

The position of C with respect to A is indicated by the vector $\overline{AC}$, and by the last remark the value of $\overline{AC}$ must be deducible from those of $\overline{AB}$ and $\overline{BC}$.

The result of the operation $\overline{AC}$ is to carry the tracing point from A to C. But the result is the same if the tracing point is carried first from A to B and then from B to C, and this is the sum of the operations $\overline{AB} + \overline{BC}$.

10. Addition of Vectors

Hence the rule for the addition of vectors may be stated thus:—From any point as origin draw the successive vectors in series, so that each vector begins at the end of the preceding one. The straight line from the origin to the extremity of the series represents the vector which is the sum of the vectors.

* A force is more completely specified as a vector localised in its line of action, called by Clifford a rotor; moreover it is only when the body on which it acts is treated as rigid that the point of application is inessential.

The order of addition is indifferent, for if we write $\overline{BC} + \overline{AB}$ the operation indicated may be performed by drawing $\overline{AD}$ parallel and equal to $\overline{BC}$, and then joining $\overline{DC}$, which, by Euclid, I. 33, is parallel and equal to $\overline{AB}$, so that by these two operations we arrive at the point C in whichever order we perform them.

The same is true for any number of vectors, take them in what order we please.

11. Subtraction of one Vector from another

To express the position of C with respect to B in terms of the positions of B and C with respect to A, we observe that we can get from B to C either by passing along the straight line $\overline{BC}$ or by passing from B to A and then from A to C. Hence

$$\begin{aligned} \overline{BC} &= \overline{BA} + \overline{AC} \\ &= \overline{AC} + \overline{BA} \text{ since the order of addition is indifferent} \\ &= \overline{AC} - \overline{AB} \text{ since } \overline{AB} \text{ is equal and opposite to } \overline{BA}. \end{aligned}$$

Or the vector $\overline{BC}$, which expresses the position of C with respect to B, is found by subtracting the vector of B from the vector of C, these vectors being drawn to B and C respectively from any common origin A.

12. Origin of Vectors

The positions of any number of particles belonging to a material system may be defined by means of the vectors drawn to each of these particles from some one point. This point is called the origin of the vectors, or, more briefly, the Origin.

This system of vectors determines the configuration of the whole system; for if we wish to know the position of any point B with respect to any other point A, it may be found from the vectors $\overline{OA}$ and $\overline{OB}$ by the equation

$$\overline{AB} = \overline{OB} - \overline{OA}.$$

We may choose any point whatever for the origin, and there is for the present no reason why we should choose one point rather than another. The configuration of the system—that is to say, the position of its parts with respect to each other—remains the same, whatever point be chosen as origin. Many inquiries, however, are simplified by a proper selection of the origin.

13. Relative Position of Two Systems

If the configurations of two different systems are known, each system having its own origin, and if we then wish to include both systems in a larger system, having, say, the same origin as the first of the two systems, we must ascertain the position of the origin of the second system with respect to that of the first, and we must be able to draw lines in the second system parallel to those in the first.

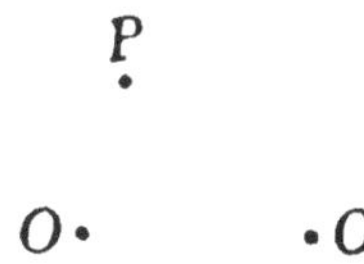

Fig. 2.

Then by Article 9 the position of a point P of the second system, with respect to the first origin, O, is represented by the sum of the vector $O'P$ of that point with respect to the second origin, O', and the vector OO' of the second origin, O', with respect to the first, O.

14. Three Data for the Comparison of Two Systems

We have an instance of this formation of a large system out of two or more smaller systems, when two neighbouring nations, having each surveyed and mapped its own territory, agree to connect their surveys so as to include both countries in one system. For this purpose three things are necessary.

1st. A comparison of the origin selected by the one country with that selected by the other.

2nd. A comparison of the directions of reference used in the two countries.

3rd. A comparison of the standards of length used in the two countries.

1. In civilised countries latitude is always reckoned from the equator, but longitude is reckoned from an arbitrary point, as Greenwich or Paris. Therefore, to make the map of Britain fit that of France, we must ascertain the difference of longitude between the Observatory of Greenwich and that of Paris.

2. When a survey has been made without astronomical instruments, the directions of reference have sometimes been those given by the magnetic compass. This was, I believe, the case in the original surveys of some of the West India islands. The results of this survey, though giving correctly the local configuration of the island, could not be made to fit properly into a general map of the world till the deviation of the magnet from the true north at the time of the survey was ascertained.

3. To compare the survey of France with that of Britain, the metre, which is the French standard of length, must be compared with the yard, which is the British standard of length.

The yard is defined by Act of Parliament 18 and 19 Vict. c. 72, July 30, 1855, which enacts "that the straight line or distance between the centres of the transverse lines in the two gold plugs in the bronze bar deposited in the office of the Exchequer shall be the genuine standard yard at 62° Fahrenheit, and if lost, it shall be replaced by means of its copies."

The metre derives its authority from a law of the French Republic in 1795. It is defined to be the distance between the ends of a certain rod of platinum made by Borda, the rod being at the temperature of melting ice. It has been found by the measurements of Captain Clarke that the metre is equal to 39·37043 British inches.

15. On the Idea of Space*

We have now gone through most of the things to be attended to with respect to the configuration of a material system. There remain, however, a few points relating to the metaphysics of the subject, which have a very important bearing on physics.

We have described the method of combining several configurations into one system which includes them all. In this way we add to the small region which we can explore by stretching our limbs the more distant regions which we can reach by walking or by being carried. To these we add those of which we learn by the reports of others, and those inaccessible regions whose positions we ascertain only by a process of calculation, till at last we recognise that every place has a definite position with respect to every other place, whether the one place is accessible from the other or not.

Thus from measurements made on the earth's surface we deduce the position of the centre of the earth relative to known objects, and we calculate the number of cubic miles in the earth's volume quite independently of any hypothesis as to what may exist at the centre of the earth, or in any other place beneath that thin layer of the crust of the earth which alone we can directly explore.

16. Error of Descartes

It appears, then, that the distance between one thing and another does not depend on any material thing between them, as Descartes seems to assert when he says (Princip. Phil., II. 18) that if that which is in a hollow vessel were taken out of it without anything

* Following Newton's method of exposition in the *Principia*, a space is assumed and a flux of time is assumed, forming together a framework into which the dynamical explanation of phenomena is set. It is part of the problem of physical astronomy to test this assumption, and to determine this frame with increasing precision. Its philosophical basis can be regarded as a different subject, to which the recent discussions on relativity as regards space and time would be attached. See Appendix I.

entering to fill its place, the sides of the vessel, having nothing between them, would be in contact.

This assertion is grounded on the dogma of Descartes, that the extension in length, breadth, and depth which constitute space is the sole essential property of matter. "The nature of matter," he tells us, "or of body considered generally, does not consist in a thing being hard, or heavy, or coloured, but only in its being extended in length, breadth, and depth" (Princip., II. 4). By thus confounding the properties of matter with those of space, he arrives at the logical conclusion that if the matter within a vessel could be entirely removed, the space within the vessel would no longer exist. In fact he assumes that all space must be always full of matter.

I have referred to this opinion of Descartes in order to show the importance of sound views in elementary dynamics. The primary property of matter was indeed distinctly announced by Descartes in what he calls the "First Law of Nature" (Princip., II. 37): "That every individual thing, so far as in it lies, perseveres in the same state, whether of motion or of rest."*

We shall see when we come to Newton's laws of motion that in the words "so far as in it lies," properly understood, is to be found the true primary definition of matter, and the true measure of its quantity. Descartes, however, never attained to a full understanding of his own words (*quantum in se est*), and so fell back on his original confusion of matter with space —space being, according to him, the only form of substance, and all existing things but affections of space. This error† runs through every part of Descartes' great work, and it forms one of the ultimate foundations of the system of Spinoza. I shall not attempt to trace it down to more modern times, but I would advise

* Compare the idea of Least Action: Appendix II.

† Some recent forms of relativity have come back to his ideas. Cf. p. 140.

those who study any system of metaphysics to examine carefully that part of it which deals with physical ideas.

We shall find it more conducive to scientific progress to recognise, with Newton, the ideas of time and space as distinct, at least in thought, from that of the material system whose relations these ideas serve to co-ordinate*.

17. On the Idea of Time

The idea of Time in its most primitive form is probably the recognition of an order of sequence in our states of consciousness. If my memory were perfect, I might be able to refer every event within my own experience to its proper place in a chronological series. But it would be difficult, if not impossible, for me to compare the interval between one pair of events and that between another pair—to ascertain, for instance, whether the time during which I can work without feeling tired is greater or less now than when I first began to study. By our intercourse with other persons, and by our experience of natural processes which go on in a uniform or a rhythmical manner, we come to recognise the possibility of arranging a system of chronology in which all events whatever, whether relating to ourselves or to others, must find their places. Of any two events, say the actual disturbance at the star in Corona Borealis, which caused the luminous effects examined spectroscopically by Mr Huggins on the 16th May, 1866, and the mental suggestion which first led Professor Adams or M. Leverrier to begin the researches which led to the discovery, by Dr Galle, on the 23rd September, 1846, of the planet Neptune, the first named must have occurred either before or after the other, or else at the same time.

Absolute, true, and mathematical Time is conceived by Newton as flowing at a constant rate, unaffected by the speed or slowness of the motions of material things.

* See Appendix I.

It is also called Duration. Relative, apparent, and common time is duration as estimated by the motion of bodies, as by days, months, and years. These measures of time may be regarded as provisional, for the progress of astronomy has taught us to measure the inequality in the lengths of days, months, and years, and thereby to reduce the apparent time to a more uniform scale, called Mean Solar Time.

18. Absolute Space

Absolute space is conceived as remaining always similar to itself and immovable. The arrangement of the parts of space can no more be altered than the order of the portions of time. To conceive them to move from their places is to conceive a place to move away from itself.

But as there is nothing to distinguish one portion of time from another except the different events which occur in them, so there is nothing to distinguish one part of space from another except its relation to the place of material bodies. We cannot describe the time of an event except by reference to some other event, or the place of a body except by reference to some other body. All our knowledge, both of time and place, is essentially relative*. When a man has acquired the habit of putting words together, without troubling himself to form the thoughts which ought to correspond to them, it is easy for him to frame an antithesis between this relative knowledge and a so-called absolute knowledge, and to point out our ignorance of the absolute position of a point as an instance of the limitation of our faculties. Any one, however, who will try to imagine the state of a mind conscious of knowing the absolute position of a point will ever after be content with our relative knowledge.

* The position seems to be that our knowledge is relative, but needs definite space and time as a frame for its coherent expression.

19. Statement of the General Maxim of Physical Science

There is a maxim which is often quoted, that "The same causes will always produce the same effects."

To make this maxim intelligible we must define what we mean by the same causes and the same effects, since it is manifest that no event ever happens more than once, so that the causes and effects cannot be the same in *all* respects. What is really meant is that if the causes differ only as regards the absolute time or the absolute place at which the event occurs, so likewise will the effects.

The following statement, which is equivalent to the above maxim, appears to be more definite, more explicitly connected with the ideas of space and time, and more capable of application to particular cases:

"The difference between one event and another does not depend on the mere difference of the times or the places at which they occur, but only on differences in the nature, configuration, or motion of the bodies concerned."

It follows from this, that if an event has occurred at a given time and place it is possible for an event exactly similar to occur at any other time and place.

There is another maxim which must not be confounded with that quoted at the beginning of this article, which asserts "That like causes produce like effects."

This is only true when small variations in the initial circumstances produce only small variations in the final state of the system*. In a great many physical phenomena this condition is satisfied; but there are other

* This implies that it is only in so far as stability subsists that principles of natural law can be formulated: it thus perhaps puts a limitation on any postulate of universal physical determinacy such as Laplace was credited with.

cases in which a small initial variation may produce a very great change in the final state of the system, as when the displacement of the "points" causes a railway train to run into another instead of keeping its proper course*.

* We may perhaps say that the observable regularities of nature belong to statistical molecular phenomena which have settled down into permanent stable conditions. In so far as the weather may be due to an unlimited assemblage of local instabilities, it may not be amenable to a finite scheme of law at all.

CHAPTER II

ON MOTION

20. Definition of Displacement

We have already compared the position of different points of a system at the same instant of time. We have next to compare the position of a point at a given instant with its position at a former instant, called the Epoch.

The vector which indicates the final position of a point with respect to its position at the epoch is called the Displacement of that point. Thus if A_1 is the initial and A_2 the final position of the point A, the line $\overline{A_1A_2}$ is the displacement of A, and any vector $\overline{oa}$ drawn from the origin o parallel and equal to $\overline{A_1A_2}$ indicates this displacement.

21. Diagram of Displacement

If another point of the system is displaced from B_1 to B_2 the vector $\overline{ob}$ parallel and equal to $\overline{B_1B_2}$ indicates the displacement of B.

In like manner the displacement of any number of points may be represented by vectors drawn from the same origin o. This system of vectors is called the Diagram of Displacement. It is not necessary to draw actual lines to represent these vectors; it is sufficient to indicate the points a, b, etc., at the extremities of the vectors. The

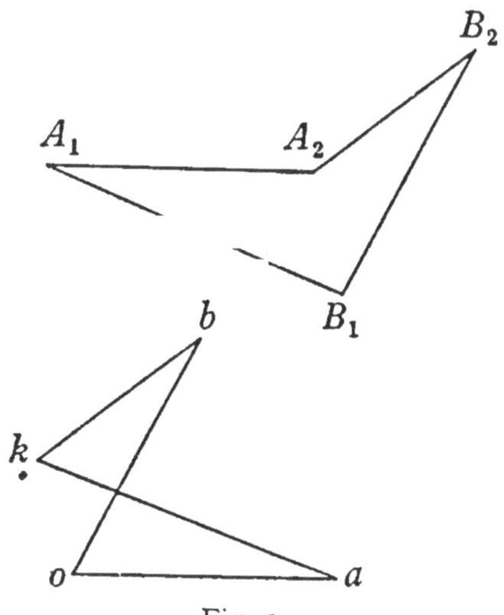

Fig. 3.

diagram of displacement may therefore be regarded as consisting of a number of points, a, b, etc., corresponding with the material particles, A, B, etc., belonging to the system, together with a point o, the position of which is arbitrary, and which is the assumed origin of all the vectors.

22. Relative Displacement

The line $\overline{ab}$ in the diagram of displacement represents the displacement of the point B with respect to A.

For if in the diagram of displacement (fig. 3) we draw $\overline{ak}$ parallel and equal to B_1A_1, and in the same direction, and join $\overline{kb}$, it is easy to show that $\overline{kb}$ is equal and parallel to $\overline{A_2B_2}$.

For the vector $\overline{kb}$ is the sum of the vectors $\overline{ka}$, $\overline{ao}$, and $\overline{ob}$, and $\overline{A_2B_2}$ is the sum of $\overline{A_2A_1}$, $\overline{A_1B_1}$, and $\overline{B_1B_2}$. But of these $\overline{ka}$ is the same as $\overline{A_1B_1}$, $\overline{ao}$ is the same as $\overline{A_2A_1}$, and $\overline{ob}$ is the same as $\overline{B_1B_2}$, and by Article 10 the order of summation is indifferent, so that the vector $\overline{kb}$ is the same, in direction and magnitude, as $\overline{A_2B_2}$. Now $\overline{ka}$ or $\overline{A_1B_1}$ represents the original position of B with respect to A, and $\overline{kb}$ or $\overline{A_2B_2}$ represents the final position of B with respect to A. Hence $\overline{ab}$ represents the displacement of B with respect to A, which was to be proved.

In Article 20 we purposely omitted to say whether the origin to which the original configuration was referred, and that to which the final configuration is referred, are absolutely the same point, or whether, during the displacement of the system, the origin also is displaced.

We may now, for the sake of argument, suppose that the origin is absolutely fixed, and that the displacements represented by $\overline{oa}$, $\overline{ob}$, etc., are the absolute displacements. To pass from this case to that in which the origin is displaced we have only to take A, one of the movable points, as origin. The absolute displacement of A being represented by $\overline{oa}$, the displacement of B with respect to A is represented, as we have seen, by $\overline{ab}$, and so on for any other points of the system.

The arrangement of the points a, b, etc., in the diagram of displacement is therefore the same, whether we reckon the displacements with respect to a fixed point or a displaced point; the only difference is that we adopt a different origin of vectors in the diagram of displacement, the rule being that whatever point we take, whether fixed or moving, for the origin of the diagram of configuration, we take the corresponding point as origin in the diagram of displacement. If we wish to indicate the fact that we are entirely ignorant of the absolute displacement in space of any point of the system, we may do so by constructing the diagram of displacement as a mere system of points, without indicating in any way which of them we take as the origin.

This diagram of displacement (without an origin) will then represent neither more nor less than all we can ever know about the displacement of the system. It consists simply of a number of points, a, b, c, etc., corresponding to the points A, B, C, etc., of the material system, and a vector, as $\overline{ab}$ represents the displacement of B with respect to A.

23. UNIFORM[1] DISPLACEMENT

When the displacements of all points of a material system with respect to an external point are the same in direction and magnitude, the diagram of displacement is reduced to two points—one corresponding to the external point, and the other to each and every point of the displaced system. In this case the points of the

[1] When the simultaneous values of a quantity for different bodies or places are equal, the quantity is said to be *uniformly* distributed in space.

system are not displaced with respect to one another, but only with respect to the external point.

This is the kind of displacement which occurs when a body of invariable form moves parallel to itself. It may be called uniform displacement.

24. On Motion

When the change of configuration of a system is considered with respect only to its state at the beginning and the end of the process of change, and without reference to the time during which it takes place, it is called the displacement of the system.

When we turn our attention to the process of change itself, as taking place during a certain time and in a continuous manner, the change of configuration is ascribed to the motion of the system.

25. On the Continuity of Motion

When a material particle is displaced so as to pass from one position to another, it can only do so by travelling along some course or path from the one position to the other.

At any instant during the motion the particle will be found at some one point of the path, and if we select any point of the path, the particle will pass that point once at least[1] during its motion.

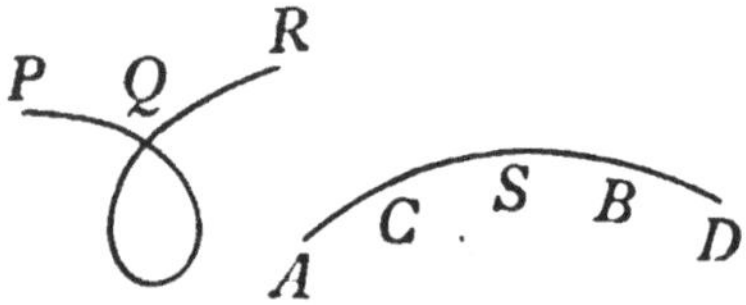

Fig. 4.

This is what is meant by saying that the particle describes a continuous path. The motion of a material particle which has continuous existence in time and space is the type and exemplar of every form of continuity.

[1] If the path cuts itself so as to form a loop, as P, *Q*, *R* (fig. 4), the particle will pass the point of intersection, *Q*, twice, and if the particle returns on its own path, as in the path *A*, *B*, C, *D*, it may pass the same point, *S*, three or more times.

26. On Constant[1] Velocity

If the motion of a particle is such that in equal intervals of time, however short, the displacements of the particle are equal and in the same direction, the particle is said to move with constant velocity.

It is manifest that in this case the path of the body will be a straight line, and the length of any part of the path will be proportional to the time of describing it.

The rate or speed of the motion is called the velocity of the particle, and its magnitude is expressed by saying that it is such a distance in such a time, as, for instance, ten miles an hour, or one metre per second. In general we select a unit of time, such as a second, and measure velocity by the distance described in unit of time.

If one metre be described in a second and if the velocity be constant, a thousandth or a millionth of a metre will be described in a thousandth or a millionth of a second. Hence, if we can observe or calculate the displacement during any interval of time, however short, we may deduce the distance which would be described in a longer time with the same velocity. This result, which enables us to state the velocity during the short interval of time, does not depend on the body's actually continuing to move at the same rate during the longer time. Thus we may know that a body is moving at the rate of ten miles an hour, though its motion at this rate may last for only the hundredth of a second.

27. On the Measurement of Velocity when Variable

When the velocity of a particle is not constant, its value at any given instant is measured by the distance which would be described in unit of time by a body having the same velocity as that which the particle has at that instant.

[1] When the successive values of a quantity for successive instants of time are equal, the quantity is said to be *constant*.

Thus when we say that at a given instant, say one second after a body has begun to fall, its velocity is 980 centimetres per second, we mean that if the velocity of a particle were constant and equal to that of the falling body at the given instant, it would describe 980 centimetres in a second.

It is specially important to understand what is meant by the velocity or rate of motion of a body, because the ideas which are suggested to our minds by considering the motion of a particle are those which Newton made use of in his method of Fluxions[1], and they lie at the foundation of the great extension of exact science which has taken place in modern times.

28. Diagram of Velocities

If the velocity of each of the bodies in the system is constant, and if we compare the configurations of the system at an interval of a unit of time, then the displacements, being those produced in unit of time in bodies moving with constant velocities, will represent those velocities according to the method of measurement described in Article 26.

If the velocities do not actually continue constant for a unit of time, then we must imagine another system consisting of the same number of bodies, and in which the velocities are the same as those of the corresponding bodies of the system at the given instant, but remain constant for a unit of time. The displacements of this system represent the velocities of the actual system at the given instant.

Another mode of obtaining the diagram of velocities of a system at a given instant is to take a small interval of time, say the *n*th part of the unit of time, so that the middle of this interval corresponds to the given

[1] According to the method of Fluxions, when the value of one quantity depends on that of another, the rate of variation of the first quantity with respect to the second may be expressed as a velocity, by imagining the first quantity to represent the displacement of a particle, while the second flows uniformly with the time.

instant. Take the diagram of displacement corresponding to this interval and magnify all its dimensions n times. The result will be a diagram of the *mean* velocities of the system during the interval. If we now suppose the number n to increase without limit the interval will diminish without limit, and the mean velocities will approximate without limit to the actual velocities at the given instant. Finally, when n becomes infinite the diagram will represent accurately the velocities at the given instant.

29. PROPERTIES OF THE DIAGRAM OF VELOCITIES (fig. 5) The diagram of velocities for a system consisting of a number of material particles consists of a number of points, each corresponding to one of the particles.

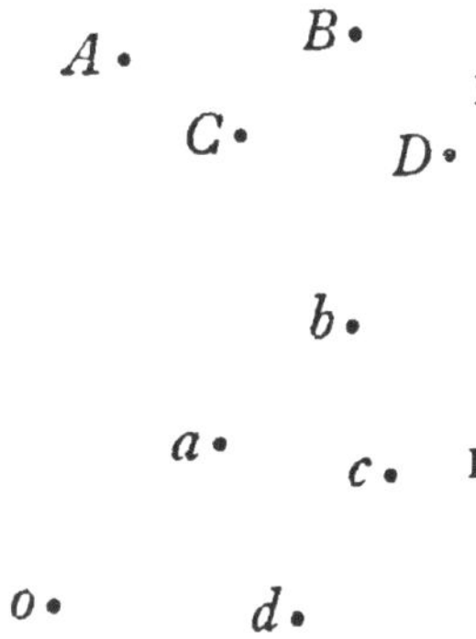

DIAGRAM OF CONFIGURATION.

DIAGRAM OF VELOCITIES.

Fig. 5.

The velocity of any particle B with respect to any other, A, is represented in direction and magnitude by the line $\overline{ab}$ in the diagram of velocities, drawn from the point a, corresponding to A, to the point b, corresponding to B.

We may in this way find, by means of the diagram, the relative velocity of any two particles. The diagram tells us nothing about the absolute velocity of any point; it expresses exactly what we can know about the motion and no more. If we choose to imagine that

$\overline{oa}$ represents the absolute velocity of A, then the absolute velocity of any other particle, B, will be represented by the vector $\overline{ob}$, drawn from o as origin to the point b, which corresponds to B.

But as it is impossible to define the position of a body except with respect to the position of some point of reference, so it is impossible to define the velocity of a body, except with respect to the velocity of the point of reference. The phrase absolute velocity has as little meaning as absolute position. It is better, therefore, not to distinguish any point in the diagram of velocities as the origin, but to regard the diagram as expressing the relations of all the velocities without defining the absolute value of any one of them.

30. Meaning of the Phrase "At Rest"

It is true that when we say that a body is at rest we use a form of words which appears to assert something about that body considered in itself, and we might imagine that the velocity of another body, if reckoned with respect to a body at rest, would be its true and only absolute velocity. But the phrase "at rest" means in ordinary language "having no velocity with respect to that on which the body stands," as, for instance, the surface of the earth or the deck of a ship. It cannot be made to mean more than this.

It is therefore unscientific to distinguish between rest and motion, as between two different states of a body in itself, since it is impossible to speak of a body being at rest or in motion except with reference, expressed or implied, to some other body.

31. On Change of Velocity

As we have compared the velocities of different bodies at the same time, so we may compare the relative velocity of one body with respect to another at different times.

If a_1, b_1, c_1, be the diagram of velocities of the system of bodies A, B, C, in its original state, and if a_2, b_2, c_2, be the diagram of velocities in the final state of the system, then if we take any point ω as origin and draw $\overline{\omega\alpha}$ equal and parallel to $\overline{a_1a_2}$, $\overline{\omega\beta}$ equal and parallel to $\overline{b_1b_2}$, $\overline{\omega\gamma}$ equal and parallel to c_1c_2, and so on, we shall form a diagram of points α, β, γ, etc., such that any line $\overline{\alpha\beta}$ in this diagram represents in direction and magnitude the change of the velocity of B with respect to A. This diagram may be called the diagram of Total Accelerations.

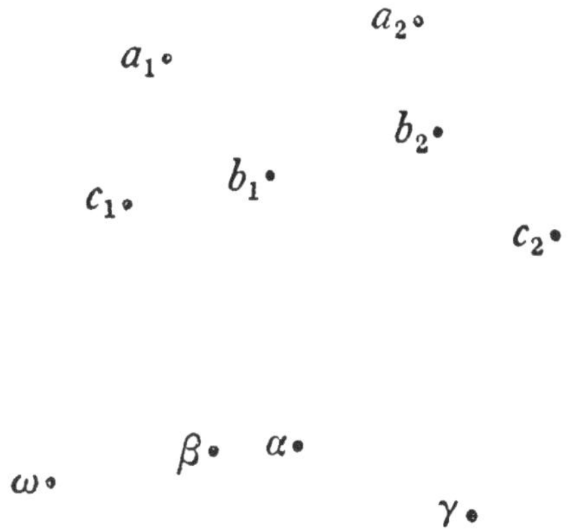

Fig. 6.

32. On Acceleration

The word Acceleration is here used to denote any change in the velocity, whether that change be an increase, a diminution, or a change of direction. Hence, instead of distinguishing, as in ordinary language, between the acceleration, the retardation, and the deflexion of the motion of a body, we say that the acceleration may be in the direction of motion, in the contrary direction, or transverse to that direction.

As the displacement of a system is defined to be the change of the configuration of the system, so the Total Acceleration of the system is defined to be the change of the velocities of the system. The process of constructing the diagram of total accelerations by a comparison of the initial and final diagrams of velocities is the same as that by which the diagram of displacement was constructed by a comparison of the initial and final diagrams of configuration.

33. On the Rate of Acceleration

We have hitherto been considering the total acceleration which takes place during a certain interval of time. If the rate of acceleration is constant, it is measured by the total acceleration in a unit of time. If the rate of acceleration is variable, its value at a given instant is measured by the total acceleration in unit of time of a point whose acceleration is constant and equal to that of the particle at the given instant.

It appears from this definition that the method of deducing the rate of acceleration from a knowledge of the total acceleration in any given time is precisely analogous to that by which the velocity at any instant is deduced from a knowledge of the displacement in any given time.

The diagram of total accelerations constructed for an interval of the nth part of the unit of time, and then magnified n times, is a diagram of the mean rates of acceleration during that interval, and by taking the interval smaller and smaller, we ultimately arrive at the true rate of acceleration at the middle of that interval.

As rates of acceleration have to be considered in physical science much more frequently than total accelerations, the word acceleration has come to be employed in the sense in which we have hitherto used the phrase rate of acceleration.

In future, therefore, when we use the word acceleration without qualification, we mean what we have here described as the rate of acceleration.

34. Diagram of Accelerations

The diagram of accelerations is a system of points, each of which corresponds to one of the bodies of the material system, and is such that any line $\overline{\alpha\beta}$ in the diagram represents the rate of acceleration of the body *B* with respect to the body *A*.

It may be well to observe here that in the diagram of configuration we use the capital letters, *A*, *B*, C, etc., to indicate the relative position of the bodies of the system; in the diagram of velocities we use the small letters, *a*, *b*, *c*, etc., to indicate the relative velocities of these bodies; and in the diagram of accelerations we use the Greek letters, α, β, γ, etc., to indicate their relative accelerations.

35. Acceleration a Relative Term

Acceleration, like position and velocity, is a relative term and cannot be interpreted absolutely*.

If every particle of the material universe within the reach of our means of observation were at a given instant to have its velocity altered by compounding therewith a new velocity, the same in magnitude and direction for every such particle, all the relative motions of bodies within the system would go on in a perfectly continuous manner, and neither astronomers nor physicists, though using their instruments all the while, would be able to find out that anything had happened†.

It is only if the change of motion occurs in a different manner in the different bodies of the system that any event capable of being observed takes place.

* A noteworthy case of relativity is Euler's investigation of the motion of a solid body as specified with reference to its own succession of instantaneous positions.

† This appears to be a very drastic postulate of relativity: a universal imposed acceleration can have no effect *during its occurrence* only when *all* applied forces are proportional to mass. See Appendix I.

CHAPTER III

ON FORCE

36. KINEMATICS AND KINETICS

WE have hitherto been considering the motion of a system in its purely geometrical aspect. We have shown how to study and describe the motion of such a system, however arbitrary, without taking into account any of the conditions of motion which arise from the mutual action between the bodies.

The theory of motion treated in this way is called Kinematics. When the mutual action between bodies is taken into account, the science of motion is called Kinetics, and when special attention is paid to force as the cause of motion, it is called Dynamics.

37. MUTUAL ACTION BETWEEN TWO BODIES—STRESS

The mutual action between two portions of matter receives different names according to the aspect under which it is studied, and this aspect depends on the extent of the material system which forms the subject of our attention.

If we take into account the whole phenomenon of the action between the two portions of matter, we call it Stress. This stress, according to the mode in which it acts, may be described as Attraction, Repulsion, Tension, Pressure, Shearing stress, Torsion, etc.

38. EXTERNAL FORCE

But if, as in Article 2, we confine our attention to one of the portions of matter, we see, as it were, only one side of the transaction—namely, that which affects the portion of matter under our consideration—and we call this aspect of the phenomenon, with reference to its effect, an External Force acting on that portion of matter, and with reference to its cause we call it the Action of the other portion of matter. The opposite aspect of the stress is called the Reaction on the other portion of matter.

39. Different Aspects of the same Phenomenon

In commercial affairs the same transaction between two parties is called *B*uying when we consider one party, Selling when we consider the other, and Trade when we take both parties into consideration.

The accountant who examines the records of the transaction finds that the two parties have entered it on opposite sides of their respective ledgers, and in comparing the books he must in every case bear in mind in whose interest each book is made up.

For similar reasons in dynamical investigations we must always remember which of the two bodies we are dealing with, so that we may state the forces in the interest of that body, and not set down any of the forces on the wrong side of the account.

40. Newton's Laws of Motion

External or "impressed" force considered with reference to its effect *—namely, the alteration of the motions of bodies—is completely defined and described in Newton's three laws of motion.

The first law tells us under what conditions there is no external force.

The second shows us how to measure the force when it exists.

The third compares the two aspects of the action between two bodies, as it affects the one body or the other.

* As to its nature, a stress, or balanced set of forces, is determined by the alteration of the permanent configuration of the bodies concerned, which reveals its existence and forms the basis of its statical measure; or else by some other property of matter. Cf. Art. 68.

41. The First Law of Motion

Law I.—*Every body perseveres in its state of rest or of moving uniformly in a straight line, except in so far as it is made to change that state by external forces.*

The experimental argument for the truth of this law is, that in every case in which we find an alteration of the state of motion of a body, we can trace this alteration to some action between that body and another, that is to say, to an external force. The existence of this action is indicated by its effect on the other body when the motion of that body can be observed. Thus the motion of a cannon ball is retarded, but this arises from an action between the projectile and the air which surrounds it, whereby the ball experiences a force in the direction opposite to its relative motion, while the air, pushed forward by an equal force, is itself set in motion, and constitutes what is called the *wind* of the cannon ball.

But our conviction of the truth of this law may be greatly strengthened by considering what is involved in a denial of it. Given a body in motion. At a given instant let it be left to itself and not acted on by any force. What will happen? According to Newton's law it will persevere in moving uniformly in a straight line, that is, its velocity will remain constant both in direction and magnitude.

If the velocity does not remain constant let us suppose it to vary. The change of velocity, as we saw in Article 31, must have a definite direction and magnitude. By the maxim of Article 19 this variation must be the same whatever be the time or place of the experiment. The direction of the change of motion must therefore be determined either by the direction of the motion itself, or by some direction fixed in the body.

Let us, in the first place, suppose the law to be that the velocity diminishes at a certain rate, which for the sake of the argument we may suppose so slow that by no experiments on moving bodies could we have detected the diminution of velocity in hundreds of years.

The velocity referred to in this hypothetical law can only be the velocity referred to a point absolutely at rest. For if it is a relative velocity its direction as well as its magnitude depends on the velocity of the point of reference.

If, when referred to a certain point, the body appears to be moving northward with diminishing velocity, we have only to refer it to another point moving northward with a uniform velocity greater than that of the body, and it will appear to be moving southward with increasing velocity.

Hence the hypothetical law is without meaning, unless we admit the possibility of defining absolute rest and absolute velocity*.

Even if we admit this as a possibility, the hypothetical law, if found to be true, might be interpreted, not as a contradiction of Newton's law, but as evidence of the resisting action of some medium in space.

To take another case. Suppose the law to be that a body, not acted on by any force, ceases at once to move. This is not only contradicted by experience, but it leads to a definition of absolute rest as the state which a body assumes as soon as it is freed from the action of external forces.

It may thus be shown that the denial of Newton's law is in contradiction to the only system of consistent doctrine about space and time which the human mind has been able to form†.

* An aether might do this. But even in Maxwell's aether an isolated body losing energy by radiation would suffer no change of velocity thereby.

† The argument of this section may be made more definite. It is a result of observation that the more isolated a body is from the influence of other bodies, the more nearly is its velocity constant with reference to an assignable frame of reference. A

42. On the Equilibrium of Forces

If a body moves with constant velocity in a straight line, the external forces, if any, which act on it, balance each other, or are in equilibrium.

Thus if a carriage in a railway train moves with constant velocity in a straight line, the external forces which act on it—such as the traction of the carriage in front of it pulling it forwards, the drag of that behind it, the friction of the rails, the resistance of the air acting backwards, the weight of the carriage acting downwards, and the pressure of the rails acting upwards—must exactly balance each other.

Bodies at rest with respect to the surface of the earth are really in motion, and their motion is not constant nor in a straight line. Hence the forces which act on them are not exactly balanced. The apparent weight of bodies is estimated by the upward force required to keep them at rest relatively to the earth. The apparent weight is

main problem of physical dynamics is to determine with increasing approximation a frame for which this principle holds, for all systems, with the greatest attainable precision. A frame of space and time thus determined has been called (after James Thomson) a frame of inertia. The statements in the text can be reconstructed with regard to a reference frame which is a frame of inertia. But given one frame of inertia, any other frame moving with any uniform translatory velocity with respect to it, is also a frame of inertia. Thus a first approximation for local purposes to a frame of inertia is one fixed with reference to the surrounding landscape; when the range of phenomena is widened, astronomers have to change to a frame containing the axis of the earth's diurnal rotation, and involving a definite value for the length of the sidereal day: this again has to be corrected for the very slow movement of the earth's axis that is revealed by the Precession of the Equinoxes: and so on. Such a frame of inertia represents in practical essentials the Newtonian absolute space and time: it is the simplest and most natural scheme of mapping an extension into which dynamical phenomena can be fitted. If we assume that space is occupied by a uniform static aether through whose mediation influences are transmitted from one material body to another, the properties of that medium will afford unique specification of an absolute space and time having physical properties as well as relations of extension. See Appendix I.

therefore rather less than the attraction of the earth, and makes a smaller angle with the axis of the earth, so that the combined effect of the supporting force and the earth's attraction is a force perpendicular to the earth's axis just sufficient to cause the body to keep to the circular path which it must describe if resting on the earth*.

43. Definition of Equal Times

The first law of motion, by stating under what circumstances the velocity of a moving body remains constant, supplies us with a method of defining equal intervals of time. Let the material system consist of two bodies which do not act on one another, and which are not acted on by any body external to the system. If one of these bodies is in motion with respect to the other, the relative velocity will, by the first law of motion, be constant and in a straight line.

Hence intervals of time are equal when the relative displacements during those intervals are equal†.

This might at first sight appear to be nothing more than a definition of what we mean by equal intervals of time, an expression which we have not hitherto defined at all.

But if we suppose another moving system of two bodies to exist, each of which is not acted upon by any body whatever, this second system will give us an independent method of comparing intervals of time.

The statement that equal intervals of time are those during which equal displacements occur in any such

* See end of Appendix I.

† This statement refers to the displacement of one body measured on a complete frame of reference attached to the other. It would not be true for two points moving with uniform velocities, if relative displacement meant merely change of distance between them. In fact their mutual distance undergoes acceleration at a rate varying inversely as the cube of that distance: to an observer not sensible of directions they would seem to repel each other with a force obeying that law of action.

system, is therefore equivalent to the assertion that the comparison of intervals of time leads to the same result whether we use the first system of two bodies or the second system as our time-piece.

We thus see the theoretical possibility of comparing intervals of time however distant, though it is hardly necessary to remark that the method cannot be put in practice in the neighbourhood of the earth, or any other large mass of gravitating matter.

44. The Second Law of Motion

Law II.—*Change of motion is proportional to the impressed force, and takes place in the direction in which the force is impressed.*

By motion Newton means what in modern scientific language is called Momentum, in which the quantity of matter moved is taken into account as well as the rate at which it travels.

By impressed force he means what is now called Impulse, in which the time during which the force acts is taken into account as well as the intensity of the force.

45. Definition of Equal Masses and of Equal Forces

An exposition of the law therefore involves a definition of equal quantities of matter and of equal forces.

We shall assume that it is possible to cause the force with which one body acts on another to be of the same intensity on different occasions.

If we admit the permanency of the properties of bodies this can be done. We know that a thread of caoutchouc when stretched beyond a certain length exerts a tension which increases the more the thread is elongated. On account of this property the thread is said to be elastic. When the same thread is drawn out to the same length it will, if its properties remain constant, exert the same tension. Now let one end of the thread be fastened to a body, M, not acted on by any other force than the tension of the thread, and let the other end be held in the hand and pulled in a constant direction with a force just sufficient to elongate the thread to a given length. The force acting on the body will then be of a given intensity, F. The body will acquire velocity, and at the end of a unit of time this velocity will have a certain value, V.

If the same string be fastened to another body, N, and pulled as in the former case, so that the elongation is the same as before, the force acting on the body will be the same, and if the velocity communicated to N in a unit of time is also the same, namely V, then we say of the two bodies M and N that they consist of equal quantities of matter, or, in modern language, they are equal in mass. In this way, by the use of an elastic string, we might adjust the masses of a number of bodies so as to be each equal to a standard unit of mass, such as a pound avoirdupois, which is the standard of mass in Britain.

46. Measurement of Mass

The scientific value of the dynamical method of comparing quantities of matter is best seen by comparing it with other methods in actual use.

As long as we have to do with bodies of exactly the same kind, there is no difficulty in understanding how the quantity of matter is to be measured. If equal quantities of the substance produce equal effects of any kind, we may employ these effects as measures of the quantity of the substance.

For instance, if we are dealing with sulphuric acid of uniform strength, we may estimate the quantity of a given portion of it in several different ways. We may weigh it, we may pour it into a graduated vessel, and so measure its volume, or we may ascertain how much of a standard solution of potash it will neutralise.

We might use the same methods to estimate a

quantity of nitric acid if we were dealing only with nitric acid; but if we wished to compare a quantity of nitric acid with a quantity of sulphuric acid we should obtain different results by weighing, by measuring, and by testing with an alkaline solution.

Of these three methods, that of weighing depends on the attraction between the acid and the earth, that of measuring depends on the volume which the acid occupies, and that of titration depends on its power of combining with potash.

In abstract dynamics, however, matter is considered under no other aspect than as that which can have its motion changed by the application of force. Hence any two bodies are of equal mass if equal forces applied to these bodies produce, in equal times, equal changes of velocity. This is the only definition of equal masses which can be admitted in dynamics, and it is applicable to all material bodies, whatever they may be made of.

It is an observed fact that bodies of equal mass, placed in the same position relative to the earth, are attracted equally towards the earth, whatever they are made of; but this is not a doctrine of abstract dynamics, founded on axiomatic principles, but a fact discovered by observation, and verified by the careful experiments of Newton*, on the times of oscillation of hollow wooden balls suspended by strings of the same length, and containing gold, silver, lead, glass, sand, common salt, wood, water, and wheat.

The fact, however, that in the same geographical position the weights of equal masses are equal, is so well established, that no other mode of comparing masses than that of comparing their weights is ever made use of, either in commerce or in science, except in researches undertaken for the special purpose of

* *Principia*, III. Prop. 6. Actual weight is a compound effect, in the main attraction, but diminished by reaction against centripetal acceleration of the mass due to the earth's rotation. See p. 143.

determining in absolute measure the weight of unit of mass at different parts of the earth's surface. The method employed in these researches is essentially the same as that of Newton, namely, by measuring the length of a pendulum which swings seconds.

The unit of mass in this country is defined by the Act of Parliament (18 & 19 Vict. c. 72, July 30, 1855) to be a piece of platinum marked "P.S., 1844, 1 lb." deposited in the office of the Exchequer, which "shall be and be denominated the Imperial Standard Pound Avoirdupois." One seven-thousandth part of this pound is a grain. The French standard of mass is the "Kilogramme des Archives," made of platinum by Borda. Professor Miller finds the kilogramme equal to 15432·34874 grains.

47. Numerical Measurement of Force

The unit of force is that force which, acting on the unit of mass for the unit of time, generates unit of velocity.

Thus the weight of a gramme—that is to say, the force which causes it to fall—may be ascertained by letting it fall freely. At the end of one second its velocity will be about 981 centimetres per second if the experiment be in Britain. Hence the weight of a gramme is represented by the number 981, if the centimetre, the gramme, and the second are taken as the fundamental units.

It is sometimes convenient to compare forces with the weight of a body, and to speak of a force of so many pounds weight or grammes weight. This is called Gravitation measure. We must remember, however, that though a pound or a gramme is the same all over the world, the weight of a pound or a gramme is greater in high latitudes than near the equator, and therefore a measurement of force in gravitation measure is of no scientific value unless it is stated in what part of the world the measurement was made.

If, as in Britain, the units of length, mass, and time are one foot, one pound, and one second, the unit of force is that which, in one second, would communicate to one pound a velocity of one foot per second. This unit of force is called a *Poundal*.

In the French metric system the units are one centimetre, one gramme, and one second. The force which in one second would communicate to one gramme a velocity of one centimetre per second is called a *Dyne*.

Since the foot is 30·4797 centimetres and the pound is 453·59 grammes, the poundal is 13825·38 dynes.

48. Simultaneous Action of Forces on a Body

Now let a unit of force act for unit of time upon unit of mass. The velocity of the mass will be changed, and the total acceleration will be unity in the direction of the force.

The magnitude and direction of this total acceleration will be the same whether the body is originally at rest or in motion*. For the expression "at rest" has no scientific meaning, and the expression "in motion," if it refers to relative motion, may mean anything, and if it refers to absolute motion can only refer to some medium fixed in space. To discover the existence of a medium, and to determine our velocity with respect to it by observation on the motion of bodies, is a legitimate scientific inquiry, but supposing all this done we should have discovered, not an error in the laws of motion, but a new fact in science.

Hence the effect of a given force on a body does not depend on the motion of that body.

Neither is it affected by the simultaneous action of other forces on the body. For the effect of these forces on the body is only to produce motion in the body, and this does not affect the acceleration produced by the first force.

* Cf. Appendix I.

Hence we arrive at the following form of the law. *When any number of forces act on a body, the acceleration due to each force is the same in direction and magnitude as if the others had not been in action.*

When a force, constant in direction and magnitude, acts on a body, the total acceleration is proportional to the interval of time during which the force acts.

For if the force produces a certain total acceleration in a given interval of time, it will produce an equal total acceleration in the next, because the effect of the force does not depend upon the velocity which the body has when the force acts on it. Hence in every equal interval of time there will be an equal change of the velocity, and the total change of velocity from the beginning of the motion will be proportional to the time of action of the force.

The total acceleration in a given time is proportional to the force.

For if several equal forces act in the same direction on the same body in the same direction, each produces its effect independently of the others. Hence the total acceleration is proportional to the number of the equal forces.

49. On Impulse

The total effect of a force in communicating velocity to a body is therefore proportional to the force and to the time during which it acts conjointly.

The product of the time of action of a force into its intensity if it is constant, or its mean intensity if it is variable, is called the *Impulse* of the force.

There are certain cases in which a force acts for so short a time that it is difficult to estimate either its intensity or the time during which it acts. But it is comparatively easy to measure the effect of the force in altering the motion of the body on which it acts, which, as we have seen, depends on the impulse.

The word impulse was originally used to denote the

effect of a force of short duration, such as that of a hammer striking a nail. There is no essential difference, however, between this case and any other case of the action of force. We shall therefore use the word impulse as above defined, without restricting it to cases in which the action is of an exceptionally transient character.

50. Relation between Force and Mass

If a force acts on a unit of mass for a certain interval of time, the impulse, as we have seen, is measured by the velocity generated.

If a number of equal forces act in the same direction, each on a unit of mass, the different masses will all move in the same manner, and may be joined together into one body without altering the phenomenon. The velocity of the whole body is equal to that produced by one of the forces acting on a unit of mass.

Hence the force required to produce a given change of velocity in a given time is proportional to the number of units of mass* of which the body consists.

51. On Momentum

The numerical value of the Momentum of a body is the product of the number of units of mass in the body into the number of units of velocity with which it is moving.

The momentum of any body is thus measured in terms of the momentum of unit of mass moving with unit of velocity, which is taken as the unit of momentum.

The direction of the momentum is the same as that of the velocity, and as the velocity can only be estimated with respect to some point of reference, so the particular value of the momentum depends on the point of refer-

* Here mass means the measure of the inertia rather than the quantity of matter; at extremely great speeds they would not be proportional, but connected by a law involving the speed, so that momentum or impulse would then be the primary quantity and inertia a derived one.

ence which we assume. The momentum of the moon, for example, will be very different according as we take the earth or the sun for the point of reference.

52. Statement of the Second Law of Motion in Terms of Impulse and Momentum

The change of momentum of a body is numerically equal to the impulse which produces it, and is in the same direction.

53. Addition of Forces

If any number of forces act simultaneously on a body, each force produces an acceleration proportional to its own magnitude (Article 48). Hence if in the diagram of accelerations (Article 34) we draw from any origin a line representing in direction and magnitude the acceleration due to one of the forces, and from the end of this line another representing the acceleration due to another force, and so on, drawing lines for each of the forces taken in any order, then the line drawn from the origin to the extremity of the last of the lines will represent the acceleration due to the combined action of all the forces.

Since in this diagram lines which represent the accelerations are in the same proportion as the forces to which these accelerations are due, we may consider the lines as representing these forces themselves. The diagram, thus understood, may be called a Diagram of Forces, and the line from the origin to the extremity of the series represents the Resultant Force.

An important case is that in which the set of lines representing the forces terminate at the origin so as to form a closed figure. In this case there is no resultant force, and no acceleration. The effects of the forces are exactly balanced, and the case is one of equilibrium. The discussion of cases of equilibrium forms the subject of the science of Statics.

It is manifest that since the system of forces is

exactly balanced, and is equivalent to no force at all*, the forces will also be balanced if they act in the same way on any other material system†, whatever be the mass of that system. This is the reason why the consideration of mass does not enter into statical investigations.

54. THE THIRD LAW OF MOTION

Law III.—*Reaction is always equal and opposite to action, that is to say, the actions of two bodies upon each other are always equal and in opposite directions.*

When the bodies between which the action takes place are not acted on by any other force, the changes in their respective momenta produced by the action are equal and in opposite directions.

The changes in the velocities of the two bodies are also in opposite directions, but not equal, except in the case of equal masses. In other cases the changes of velocity are in the inverse ratio of the masses.

55. ACTION AND REACTION ARE THE PARTIAL ASPECTS OF A STRESS

We have already (Article 37) used the word Stress to denote the mutual action between two portions of matter. This word was borrowed from common language, and invested with a precise scientific meaning by the late Professor Rankine, to whom we are indebted for several other valuable scientific terms.

As soon as we have formed for ourselves the idea of a stress, such as the Tension of a rope or the Pressure between two bodies, and have recognised its double aspect as it affects the two portions of matter between

* Except however as regards the strains which the system of forces sets up in a deformable body, in cases when they do not all act at the same point. It is when these strains are not regarded, or the body on which they act is considered as perfectly rigid, that we can speak of the statical equivalence of two systems of forces.

† If the forces do not act at the same point, the system must be a rigid one, else it will be deformed by them.

which it acts, the third law of motion is seen to be equivalent to the statement that all force is of the nature of stress, that stress exists only between two portions of matter, and that its effects on these portions of matter (measured by the momentum generated in a given time) are equal and opposite.

The stress is measured numerically by the force exerted on either of the two portions of matter. It is distinguished as a tension when the force acting on either portion is towards the other, and as a pressure when the force acting on either portion is away from the other.

When the force is inclined to the surface which separates the two portions of matter the stress cannot be distinguished by any term in ordinary language, but must be defined by technical mathematical terms.

When a tension is exerted between two bodies by the medium of a string, the stress, properly speaking, is between any two parts into which the string may be supposed to be divided by an imaginary section or transverse interface. If, however, we neglect the weight of the string, each portion of the string is in equilibrium under the action of the tensions at its extremities, so that the tensions at any two transverse interfaces of the string must be the same. For this reason we often speak of the tension of the string as a whole, without specifying any particular section of it, and also the tension between the two bodies, without considering the nature of the string through which the tension is exerted.

56. Attraction and Repulsion

There are other cases in which two bodies at a distance appear mutually to act on each other, though we are not able to detect any intermediate body, like the string in the former example, through which the action takes place. For instance, two magnets or two electrified bodies appear to act on each other when placed at considerable distances apart, and the motions of the heavenly bodies are observed to be affected in a manner which depends on their relative position.

This mutual action between distant bodies is called attraction when it tends to bring them nearer, and repulsion when it tends to separate them.

In all cases, however, the action and reaction between the bodies are equal and opposite.

57. The Third Law true of Action at a Distance

The fact that a magnet draws iron towards it was noticed by the ancients, but no attention was paid to the force with which the iron attracts the magnet. Newton, however, by placing the magnet in one vessel and the iron in another, and floating both vessels in water so as to touch each other, showed experimentally that as neither vessel was able to propel the other along with itself through the water, the attraction of the iron on the magnet must be equal and opposite to that of the magnet on the iron, both being equal to the pressure between the two vessels.

Having given this experimental illustration Newton goes on to point out the consequence of denying the truth of this law. For instance, if the attraction of any part of the earth, say a mountain, upon the remainder of the earth were greater or less than that of the remainder of the earth upon the mountain, there would be a residual force, acting upon the system of the earth and the mountain as a whole, which would cause it to move off, with an ever-increasing velocity, through infinite space.

58. Newton's Proof not Experimental

This is contrary to the first law of motion, which asserts that a body does not change its state of motion unless acted on by *external* force. It cannot be affirmed to be contrary to experience, for the effect of an inequality between the attraction of the earth on the mountain and the mountain on the earth would be the same as that of a force equal to the difference of these attractions acting in the direction of the line joining the centre of the earth with the mountain.

If the mountain were at the equator the earth would be made to rotate about an axis parallel to the axis about which it would otherwise rotate, but not passing exactly through the centre of the earth's mass*.

If the mountain were at one of the poles, the constant force parallel to the earth's axis would cause the orbit of the earth about the sun to be slightly shifted to the north or south of a plane passing through the centre of the sun's mass.

If the mountain were at any other part of the earth's surface its effect would be partly of the one kind and partly of the other.

Neither of these effects, unless they were very large, could be detected by direct astronomical observations, and the indirect method of detecting small forces, by their effect in slowly altering the elements of a planet's orbit, presupposes that the law of gravitation is known to be true. To prove the laws of motion by the law of gravitation would be an inversion of scientific order. We might as well prove the law of addition of numbers by the differential calculus.

We cannot, therefore, regard Newton's statement as an appeal to experience and observation, but rather as a deduction of the third law of motion from the first.

* This is because such a residual force would revolve along with the earth's diurnal motion. If F is this force, E the earth's mass and ω its angular velocity, the altered axis of rotation would be at a distance R from the centre of mass such that $F=E\omega^2R$.

In the next sentence the direction of the residual force is constant; and the earth being held in an orbit around the sun by the gravitational attraction, that force is transferred to the solar system as a whole, to which accordingly, and not to the earth alone, the final statement in Art. 57 would apply.

CHAPTER IV

ON THE PROPERTIES OF THE CENTRE OF MASS OF A MATERIAL SYSTEM

59. Definition of a Mass-Vector

We have seen that a vector represents the operation of carrying a tracing point from a given origin to a given point.

Let us define a mass-vector as the operation of carrying a given mass from the origin to the given point. The direction of the mass-vector is the same as that of the vector of the mass, but its magnitude is the product of the mass into the vector of the mass.

Thus if $\overline{OA}$ is the vector of the mass A, the mass-vector is $\overline{OA} \, . \, A$.

60. Centre of Mass of Two Particles

If A and B are two masses, and if a point C be taken in the straight line $\overline{AB}$, so that $\overline{BC}$ is to $\overline{CA}$ as A to B, then the mass-vector of a mass $A + B$ placed at C is equal to the sum of the mass-vectors of A and B.

$$\text{For } \overline{OA} \, . \, A + \overline{OB} \, . \, B = (\overline{OC} + \overline{CA}) \, A + (\overline{OC} + \overline{CB}) \, B$$
$$= \overline{OC} \, (A + B) + \overline{CA} \, . \, A + \overline{CB} \, . \, B.$$

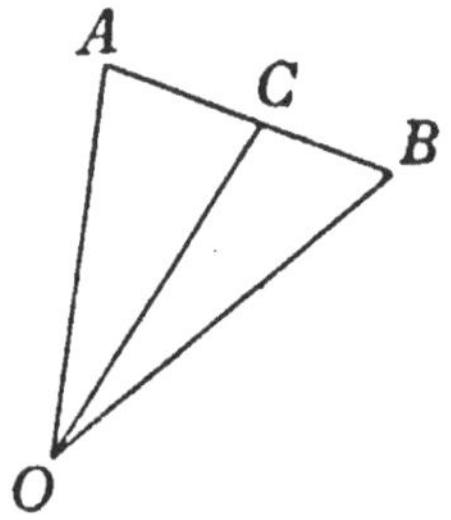

Fig. 7.

Now the mass-vectors $\overline{CA} \, . \, A$ and $\overline{CB} \, . \, B$ are equal and opposite, and so destroy each other, so that

$$\overline{OA} \, . \, A + \overline{OB} \, . \, B = \overline{OC} \, (A + B)$$

or, C is a point such that if the masses of A and B were coneentrated at C, their mass-vector from any origin O would be the same as when A and B are in their actual positions. The point C is called the *Centre of Mass* of A and B.

61. Centre of Mass of a System

If the system consists of any number of particles, we may begin by finding the centre of mass of any two particles, and substituting for the two particles a particle equal to their sum placed at their centre of mass. We may then find the centre of mass of this particle, together with the third particle of the system, and place the sum of the three particles at this point, and so on till we have found the centre of mass of the whole system.

The mass-vector drawn from any origin to a mass equal to that of the whole system placed at the centre of mass of the system is equal to the sum of the mass-vectors drawn from the same origin to all the particles of the system.

It follows, from the proof in Article 60, that the point found by the construction here given satisfies this condition. It is plain from the condition itself that only one point can satisfy it. Hence the construction must lead to the same result, as to the position of the centre of mass, in whatever order we take the particles of the system.

The centre of mass is therefore a definite point in the diagram of the configuration of the system. By assigning to the different points in the diagrams of displacement, velocity, total acceleration, and rate of acceleration, the masses of the bodies to which they correspond, we may find in each of these diagrams a point which corresponds to the centre of mass, and indicates the displacement, velocity, total acceleration, or rate of acceleration of the centre of mass.

62. Momentum represented as the Rate of Change of a Mass-Vector

In the diagram of velocities, if the points o, a, b, c, correspond to the velocities of the origin O and the bodies A, B, C, and if p be the centre of mass of A

and B placed at a and b respectively, and if q is the centre of mass of $A + B$ placed at p and C at c, then q will be the centre of mass of the system of bodies A, B, C, at a, b, c, respectively.

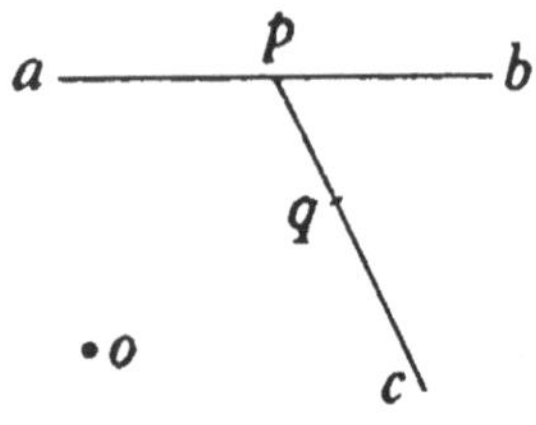

Fig. 8.

The velocity of A with respect to O is indicated by the vector $\overline{oa}$, and that of B and C by $\overline{ob}$ and $\overline{oc}$. $\overline{op}$ is the velocity of the centre of mass of A and B, and $\overline{oq}$ that of the centre of mass of A, B, and C, with respect to O.

The momentum of A with respect to O is the product of the velocity into the mass, or $\overline{oa} \,.\, A$, or what we have already called the mass-vector, drawn from o to the mass A at a. Similarly the momentum of any other body is the mass-vector drawn from o to the point on the diagram of velocities corresponding to that body, and the momentum of the mass of the system concentrated at the centre of mass is the mass-vector drawn from o to the whole mass at q.

Since, therefore, a mass-vector in the diagram of velocities is what we have already defined as a momentum, we may state the property proved in Article 61 in terms of momenta, thus: The momentum of a mass equal to that of the whole system, moving with the velocity of the centre of mass of the system, is equal in magnitude and parallel in direction to the sum of the momenta of all the particles of the system.

63. Effect of External Forces on the Motion of the Centre of Mass

In the same way in the diagram of Total Acceleration the vectors $\overline{\omega a}$, $\overline{\omega \beta}$, etc., drawn from the origin, represent the change of velocity of the bodies A, B, etc., during a certain interval of time. The corresponding mass-vectors, $\overline{\omega a} \,.\, A$, $\overline{\omega \beta} \,.\, B$, etc., represent the corresponding changes of momentum, or, by the second law of motion, the impulses of the forces acting on these bodies during that interval of time. If κ is the centre of mass of the system, $\overline{\omega\kappa}$ is the change of velocity during the interval, and $\overline{\omega\kappa}\,(A + B + C)$ is the momentum generated in the mass concentrated at the centre of gravity. Hence, by Article 61, the change of momentum of the imaginary mass equal to that of the whole system concentrated at the centre of mass is equal to the sum of the changes of momentum of all the different bodies of the system.

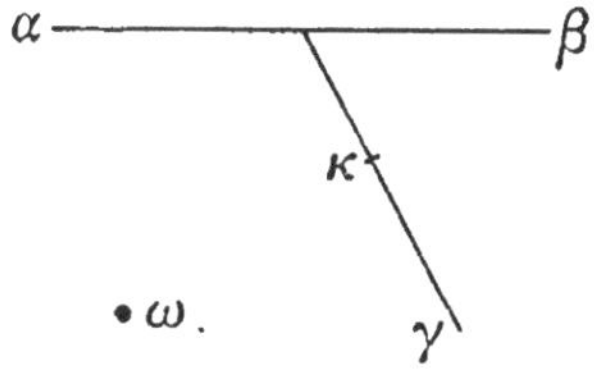

Fig. 9.

In virtue of the second law of motion we may put this result in the following form:

The effect of the forces acting on the different bodies of the system in altering the motion of the centre of mass of the system is the same as if all these forces had been applied to a mass equal to the whole mass of the system, and coinciding with its centre of mass.

64. The Motion of the Centre of Mass of a System is not affected by the Mutual Action of the Parts of the System

For if there is an action between two parts of the system, say A and B, the action of A on B is always, by the third law of motion, equal and opposite to the reaction of B on A. The momentum generated in B by the action of A during any interval is therefore equal and opposite to that generated in A by the reaction of B during the same interval, and the motion of the centre of mass of A and B is therefore not affected by their mutual action.

We may apply the result of the last article to this case and say, that since the forces on A and on B arising from their mutual action are equal and opposite, and since the effect of these forces on the motion of the centre of mass of the system is the same as if they had been applied to a particle whose mass is equal to the whole mass of the system, and since the effect of two forces equal and opposite to each other is zero, the motion of the centre of mass will not be affected.

65. First and Second Laws of Motion

This is a very important result. It enables us to render more precise the enunciation of the first and second laws of motion, by defining that by the velocity of a body is meant the velocity of its centre of mass. The body may be rotating, or it may consist of parts, and be capable of changes of configuration, so that the motions of different parts may be different, but we can still assert the laws of motion in the following form:

Law I.—The centre of mass of the system perseveres in its state of rest, or of uniform motion in a straight line, except in so far as it is made to change that state by forces acting on the system from without.

Law II.—The change of momentum* of the system during any interval of time is measured by the sum of the impulses of the external forces during that interval.

66. Method of treating Systems of Molecules

When the system is made up of parts which are so small that we cannot observe them, and whose motions are so rapid and so variable that even if we could observe them we could not describe them, we are still able to deal with the motion of the centre of mass of the system, because the internal forces which cause the variation of the motion of the parts do not affect the motion of the centre of mass.

* Meaning in the present connexion momentum of translatory motion or linear momentum, as distinguished from the angular momentum of rotatory motion. Cf. Art. 69. The law holds in an extended sense for both together. Cf. Art. 70.

67. By the Introduction of the Idea of Mass we pass from Point-Vectors, Point Displacements, Velocities, Total Accelerations, and Rates of Acceleration, to Mass-Vectors, Mass Displacements, Momenta, Impulses, and Moving Forces.

In the diagram of rates of acceleration (Fig. 9, Article 63) the vectors $\overline{\omega\alpha}$, $\overline{\omega\beta}$, etc., drawn from the origin, represent the rates of acceleration of the bodies A, B, etc., at a given instant, with respect to that of the origin O.

The corresponding mass-vectors, $\overline{\omega\alpha} \,.\, A$, $\overline{\omega\beta} \,.\, B$, etc., represent the forces acting on the bodies A, B, etc.

We sometimes speak of several forces acting on a body, when the force acting on the body arises from several different causes, so that we naturally consider the parts of the force arising from these different causes separately.

But when we consider force, not with respect to its causes, but with respect to its effect—that of altering the motion of a body—we speak not of the forces, but of the force acting on the body, and this force is measured by the rate of change of the momentum of the body, and is indicated by the mass-vector in the diagram of rates of acceleration*.

* This distinction is conveniently expressed by the terms *applied forces* and *effective forces*. For a single particle these two sets are statically equivalent. Therefore for any body which can be regarded as a system of particles held together by mutual influences, the same must be true in the aggregate, when their mutual forces are also included among the applied forces. But these internal mutual forces must in any case immediately become adjusted so as to be statically equilibrated by themselves, otherwise the parts of the body would be set by them into continually accelerated motion even when it is removed from all external influences. Therefore, leaving them out of account, the forces applied from without are statically equivalent, *as regards* the given type of body, to the effective forces that accelerate the particles or elements of mass of that body. This is the Principle of d'Alembert: though it is implied in the Newtonian scheme, being provided for by the Third Law, its more explicit recognition in 1743 gave rise to great simplification in the treatment of abstruse dynamical problems, as exemplified in d'Alembert's discussion of the spin of the earth's axis which causes the precession of the equinoxes, by reducing them to problems of statics.

We have thus a series of different kinds of mass-vectors corresponding to the series of vectors which we have already discussed.

We have, in the first place, a system of mass-vectors with a common origin, which we may regard as a method of indicating the distribution of mass in a material system, just as the corresponding system of vectors indicates the geometrical configuration of the system.

In the next place, by comparing the distribution of mass at two different epochs, we obtain a system of mass-vectors of displacement.

The rate of mass displacement is momentum, just as the rate of displacement is velocity.

The change of momentum is impulse, as the change of velocity is total acceleration.

The rate of change of momentum is moving force, as the rate of change of velocity is rate of acceleration.

68. Definition of a Mass-Area

When a material particle moves from one point to another, twice the area swept out by the vector of the particle multiplied by the mass of the particle is called the mass-area of the displacement of the particle with respect to the origin from which the vector is drawn.

If the area is in one plane, the direction of the mass-area is normal to the plane, drawn so that, looking in the positive direction along the normal, the motion of the particle round its area appears to be the direction of the motion of the hands of a watch*.

If the area is not in one plane, the path of the particle must be divided into portions so small that each coincides sensibly with a straight line, and the mass-areas corresponding to these portions must be added together by the rule for the addition of vectors.

* Stated in absolute terms, the motion round the area is in the direction of a right-handed screw motion which progresses along the normal in the positive direction.

69. Angular Momentum

The rate of change of a mass-area is twice the mass of the particle into the triangle, whose vertex is the origin and whose base is the velocity of the particle measured along the line through the particle in the direction of its motion. The direction of this mass-area is indicated by the normal drawn according to the rule given above.

The rate of change of the mass-area of a particle is called the Angular Momentum of the particle about the origin, and the sum of the angular momenta of all the particles is called the angular momentum of the system about the origin.

The angular momentum of a material system with respect to a point is, therefore, a quantity having a definite direction as well as a definite magnitude.

The definition of the angular momentum of a particle about a point may be expressed somewhat differently as the product of the momentum of the particle with respect to that point into the perpendicular from that point on the line of motion of the particle at that instant.

70. Moment of a Force about a Point

The rate of increase of the angular momentum of a particle is the continued product of the rate of acceleration of the velocity of the particle into the mass of the particle into the perpendicular from the origin on the line through the particle along which the acceleration takes place. In other words, it is the product of the moving force acting on the particle into the perpendicular from the origin on the line of action of this force.

Now the product of a force into the perpendicular from the origin on its line of action is called the Moment of the force about the origin. The axis of the moment, which indicates its direction, is a vector drawn perpendicular to the plane passing through the force and the origin, and in such a direction that, looking along this line in the direction in which it is drawn, the force tends to move the particle round the origin in the direction of the hands of a watch.

Hence the rate of change of the angular momentum of a particle about the origin is measured by the moment of the force which acts on the particle about that point.

The rate of change of the angular momentum of a material system about the origin is in like manner measured by the geometric sum of the moments of the forces which act on the particles of the system.

71. Conservation of Angular Momentum

Now consider any two particles of the system. The forces acting on these two particles, arising from their mutual action, are equal, opposite, and in the same straight line. Hence the moments of these forces about any point as origin are equal, opposite, and about the same axis. The sum of these moments is therefore zero. In like manner the mutual action between every other pair of particles in the system consists of two forces, the sum of whose moments is zero.

Hence the mutual action between the bodies of a material system does not affect the geometric sum of the moments of the forces. The only forces, therefore, which need be considered in finding the geometric sum of the moments are those which are external to the system—that is to say, between the whole or any part of the system and bodies not included in the system.

The rate of change of the angular momentum of the system is therefore measured by the geometric sum of the moments of the external forces acting on the system.

If the directions of all the external forces pass through the origin, their moments are zero, and the angular momentum of the system will remain constant.

When a planet describes an orbit about the sun, the direction of the mutual action between the two bodies always passes through their common centre of mass. Hence the angular momentum of either body about their common centre of mass remains constant, so far as these two bodies only are concerned, though it may be affected by the action of other planets. If, however, we include all the planets in the system, the geometric sum of their angular momenta about their common centre of mass will remain absolutely constant*, whatever may be their mutual actions, provided no force arising from bodies external to the whole solar system acts in an unequal manner upon the different members of the system.

* That is, the plane of the total angular momentum of the solar system is invariable in direction in space.

The plane of this resultant angular momentum, called by Laplace the "invariable plane," is fundamental for the exact specification of the motion of the solar system.

CHAPTER V

ON WORK AND ENERGY

72. Definitions

WORK *is the act of* producing a change of configuration in a system in opposition to a force which resists that change.*

ENERGY *is the capacity of doing work.*

When the nature of a material system is such that if, after the system has undergone any series of changes it is brought back in any manner to its original state, the whole work done by external agents on the system is equal to the whole work done by the system in overcoming external forces, the system is called a CONSERVATIVE SYSTEM†.

73. Principle of Conservation of Energy

The progress of physical science has led to the discovery and investigation of different forms of energy, and to the establishment of the doctrine that all material systems may be regarded as conservative systems, *provided* that all the different forms of energy which exist in these systems are taken into account.

This doctrine, considered as a deduction from observation and experiment, can, of course, assert no more than that no instance of a non-conservative system has hitherto been discovered.

As a scientific or science-producing doctrine, how-

* The work done is a quantitative measure of the effort expended in deranging the system, in terms of the consumption of energy that is required to give effect to it

The idea of work implies a fund of energy, from which the work is supplied.

† As distinguished from a system in which the energy available for work becomes gradually degraded to less available forms by frictional agencies, called a *Dissipative System*. Cf. Art. 93.

ever, it is always acquiring additional credibility from the constantly increasing number of deductions which have been drawn from it, and which are found in all cases to be verified by experiment.

In fact the doctrine of the Conservation of Energy is the one generalised statement which is found to be consistent with fact, not in one physical science only, but in all.

When once apprehended it furnishes to the physical inquirer a principle on which he may hang every known law relating to physical actions, and by which he may be put in the way to discover the relations of such actions in new branches of science*.

For such reasons the doctrine is commonly called the Principle of the Conservation of Energy.

74. GENERAL STATEMENT OF THE PRINCIPLE OF THE CONSERVATION OF ENERGY

The total energy of any material system is a quantity which can neither be increased nor diminished by any action between the parts of the system, though it may be transformed into any of the forms of which energy is susceptible.

If, by the action of some agent external to the system, the configuration of the system is changed, while the forces of the system resist this change of configuration, the external agent is said to do work on the system. In this case the energy of the system is increased by the amount of work done on it by the external agent.

If, on the contrary, the forces of the system produce a change of configuration which is resisted by the external agent, the system is said to do work on the

* Every law relating to the forces of statical or steady systems is involved implicitly in the complete expression for the Energy of the system. But in a kinetic system, where force is being used in producing energy of motion, a more elaborate principle is required, that of Least Action, for example. See *infra*, Chapter IX.

external agent, and the energy of the system is diminished by the amount of work which it does.

Work, therefore, is a transference of energy from one system to another; the system which gives out energy is said to do work on the system which receives it, and the amount of energy given out by the first system is always exactly equal to that received by the second.

If, therefore, we include both systems in one larger system, the energy of the total system is neither increased nor diminished by the action of the one partial system on the other.

75. Measurement of Work

Work done by an external agent on a material system may be described as a change* in the configuration of the system taking place under the action of an external force tending to produce that change.

Thus, if one pound is lifted one foot from the ground by a man in opposition to the force of gravity, a certain amount of work is done by the man, and this quantity is known among engineers as one foot-pound.

Here the man is the external agent, the material system consists of the earth and the pound, the change of configuration is the increase of the distance between the matter of the earth and the matter of the pound, and the force is the upward force exerted by the man in lifting the pound, which is equal and opposite to the weight of the pound. To raise the pound a foot higher would, if gravity were a uniform force, require exactly the same amount of work. It is true that gravity is not really uniform, but diminishes as we ascend from the earth's surface, so that a foot-pound is not an accurately

* See footnote, Art. 72.

These ideas, leading to an estimate of the total effect by *work* done rather than *momentum* produced, are of the kind that were enforced by Leibniz. What was then mainly needed to avoid confusion was a set of names for the different effects.

known quantity, unless we specify the intensity of gravity at the place. But for the purpose of illustration we may assume that gravity is uniform for a few feet of ascent, and in that case the work done in lifting a pound would be one foot-pound for every foot the pound is lifted.

To raise twenty pounds of water ten feet high requires 200 foot-pounds of work. To raise one pound ten feet high requires ten foot-pounds, and as there are twenty pounds the whole work is twenty times as much, or two hundred foot-pounds.

The quantity of work done is, therefore, proportional to the product of the numbers representing the force exerted and the displacement in the direction of the force.

In the case of a foot-pound the force is the weight of a pound—a quantity which, as we know, is different in different places. The weight of a pound expressed in absolute measure is numerically equal to the intensity of gravity, the quantity denoted by g, the value of which in poundals to the pound varies from 32·227 at the poles to 32·117 at the equator, and diminishes without limit as we recede from the earth. In dynes to the gramme it varies from 978·1 to 983·1. Hence, in order to express work in a uniform and consistent manner, we must multiply the number of foot-pounds by the number representing the intensity of gravity at the place. The work is thus reduced to foot-poundals. We shall always understand work to be measured in this manner and reckoned in foot-poundals when no other system of measurement is mentioned. When work is expressed in foot-pounds the system is that of *gravitation-measures*, which is not a complete system unless we also know the intensity of gravity at the place.

In the metrical system the unit of work is the Erg, which is the work done by a dyne acting through a centimetre. There are 421393·8 ergs in a foot-poundal.

76. POTENTIAL ENERGY

The work done by a man in raising a heavy body is done in overcoming the attraction between the earth and that body. The energy of the material system, consisting of the earth and the heavy body, is thereby increased. If the heavy body is the leaden weight of a clock, the energy of the clock is increased by winding it up, so that the clock is able to go for a week in spite of the friction of the wheels and the resistance of the air to the motion of the pendulum, and also to give out energy in other forms, such as the communication of the vibrations to the air, by which we hear the ticking of the clock.

When a man winds up a watch he does work in changing the form of the mainspring by coiling it up. The energy of the mainspring is thereby increased, so that as it uncoils itself it is able to keep the watch going.

In both these cases the energy communicated to the system depends upon a change of configuration.

77. KINETIC ENERGY

But in a very important class of phenomena the work is done in changing the velocity of the body on which it acts. Let us take as a simple case that of a body moving without rotation under the action of a force. Let the mass of the body be M pounds, and let a force of F poundals act on it in the line of motion during an interval of time, T seconds. Let the velocity at the beginning of the interval be V and that at the end V' feet per second, and let the distance travelled by the body during the time be S feet. The original momentum is MV, and the final momentum is MV', so that the increase of momentum is $M(V'-V)$, and this, by the second law of motion, is equal to FT, the *impulse* of the force F acting for the time T. Hence

$$FT = M(V' - V) \qquad \text{.....(1).}$$

Since the velocity increases uniformly with the time [when the force is constant], the mean velocity is the arithmetical mean of the original and final velocities, or $\frac{1}{2}(V' + V)$.

We can also determine the mean velocity by dividing the space S by the time T, during which it is described. Hence

$$\frac{S}{T} = \frac{1}{2}(V' + V) \qquad \ldots\ldots(2).$$

Multiplying the corresponding members of equations (1) and (2) each by each we obtain

$$FS = \tfrac{1}{2}MV'^2 - \tfrac{1}{2}MV^2 \qquad \ldots\ldots(3).$$

Here FS is the work done by the force F acting on the body while it moves through the space S in the direction of the force, and this is equal to the excess of $\frac{1}{2}MV'^2$ above $\frac{1}{2}MV^2$. If we call $\frac{1}{2}MV^2$, or half the product of the mass into the square of the velocity, the *kinetic energy* of the body at first, then $\frac{1}{2}MV'^2$ will be the kinetic energy after the action of the force F through the space S. The energy is here expressed in foot-poundals.

We may now express the equation in words by saying that the work done by the force F in changing the motion of the body is measured by the increase of the kinetic energy of the body during the time that the force acts.

We have proved that this is true, when the interval of time is so small that we may consider the force as constant during that time, and the mean velocity during the interval as the arithmetical mean of the velocities at the beginning and end of the interval. This assumption, which is exactly true when the force is constant, however long the interval may be, becomes in every case more and more nearly true as the interval of time taken becomes smaller and smaller. By dividing the whole time of action into small parts, and proving that in each of these the work done is equal to the increase of the kinetic energy of the body, we may, by adding the successive portions of the work and the successive increments of energy, arrive at the result that the total work done by the force is equal to the total increase of kinetic energy.

If the force acts on the body in the direction opposite to its motion, the kinetic energy of the body will be diminished instead of being increased, and the force, instead of doing work on the body, will act as a resistance, which the body, in its motion, overcomes. Hence a moving body, as long as it is in motion, can do work in overcoming resistance, and the work done by the moving body is equal to the diminution of its kinetic energy, till at last, when the body is brought to rest, its kinetic energy is exhausted, and the whole work it has done is then equal to the whole kinetic energy which it had at first.

We now see the appropriateness of the name *kinetic energy*, which we have hitherto used merely as a name to denote the product $\frac{1}{2}MV^2$. For the energy of a body has been defined as the capacity which it has of doing work, and it is measured by the work which it can do. The *kinetic* energy of a body is the energy it has in virtue of being in *motion*, and we have now shown that its value is expressed by $\frac{1}{2}MV^2$ or $\frac{1}{2}MV \times V$, that is, half the product of its momentum into its velocity.

78. Oblique Forces

If the force acts on the body at right angles to the direction of its motion it does no work on the body, and it alters the direction but not the magnitude of the velocity. The kinetic energy, therefore, which depends on the square of the velocity, remains unchanged.

If the direction of the force is neither coincident with, nor at right angles to, that of the motion of the body we may resolve the force into two components, one of which is at right angles to the direction of motion, while the other is in the direction of motion (or in the opposite direction).

The first of these components may be left out of consideration in all calculations about energy, since it neither does work on the body nor alters its kinetic energy.

The second component is that which we have already considered. When it is in the direction of motion it increases the kinetic energy of the body by the amount of work which it does on the body. When it is in the opposite direction the kinetic energy of the body is diminished by the amount of work which the body does against the force.

Hence in all cases the increase of kinetic energy is equal to the work done on the body by external agency, and the diminution of kinetic energy is equal to the work done by the body against external resistance.

79. Kinetic Energy of Two Particles referred to their Centre of Mass

The kinetic energy of a material system is equal to the kinetic energy of a mass equal to that of the system moving with the velocity of the centre of mass of the system, together with the kinetic energy due to the motion of the parts of the system relative to its centre of mass.

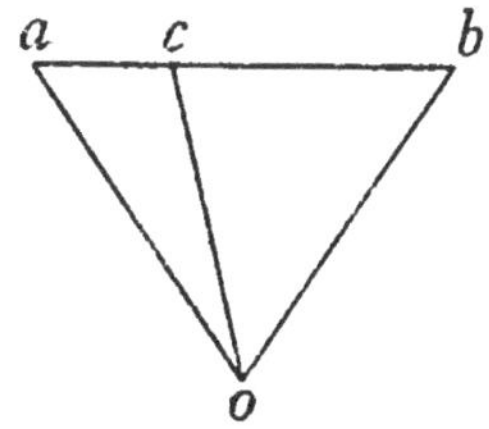

Fig. 10.

Let us begin with the case of two particles whose masses are A and B, and whose velocities are represented in the diagram of velocities by the lines oa and ob. If c is the centre of mass of a particle equal to A placed at a, and a particle equal to B placed at b, then oc will represent the velocity of the centre of mass of the two particles.

The kinetic energy of the system is the sum of the kinetic energies of the particles, or

$$T = \tfrac{1}{2}Aoa^2 + \tfrac{1}{2}Bob^2.$$

Expressing oa^2 and ob^2 in terms of oc, ca and cb, and the angle $oca = \theta$,

$$T = \tfrac{1}{2}Aoc^2 + \tfrac{1}{2}Aca^2 - Aoc \, . \, ca \cos\theta$$
$$+ \tfrac{1}{2}Boc^2 + \tfrac{1}{2}Bcb^2 - Boc \, . \, cb \cos\theta.$$

But since c is the centre of mass of A at a, and B at b,

$$Aca + Bcb = 0.$$

Hence adding

$$T = \tfrac{1}{2}(A + B)\, oc^2 + \tfrac{1}{2}Aca^2 + \tfrac{1}{2}Bcb^2,$$

or, the kinetic energy of the system of two particles A and B is equal to that of a mass equal to $(A + B)$ moving with the velocity of the centre of mass, together with that of the motion of the particles relative to the centre of mass.

80. Kinetic Energy of a Material System referred to its Centre of Mass

We have begun with the case of two particles, because the motion of a particle is assumed to be that of its centre of mass, and we have proved our proposition true for a system of two particles. But if the proposition is true for each of two material systems taken separately, it must be true of the system which they form together. For if we now suppose oa and ob to represent the velocities of the centres of mass of two material systems A and B, then oc will represent the velocity of the centre of mass of the combined system $A + B$, and if T_A represents the kinetic energy of the motion of the system A relative to its own centre of mass, and T_B the same for the system B, then if the proposition is true for the systems A and B taken separately, the kinetic energy of A is

$$\tfrac{1}{2}Aoa^2 + T_A,$$

and that of B $\quad \tfrac{1}{2}Bob^2 + T_B.$

The kinetic energy of the whole is therefore

$$\tfrac{1}{2}Aoa^2 + \tfrac{1}{2}Bob^2 + T_A + T_B,$$

or, $$\tfrac{1}{2}(A+B)\,oc^2 + \tfrac{1}{2}Aca^2 + T_A + \tfrac{1}{2}Bcb^2 + T_B.$$

The first term represents the kinetic energy of a mass equal to that of the whole system moving with the velocity of the centre of mass of the whole system.

The second and third terms, taken together, represent the kinetic energy of the system A relative to the centre of gravity of the whole system, and the fourth and fifth terms represent the same for the system B.

Hence if the proposition is true for the two systems A and B taken separately, it is true for the system compounded of A and B. But we have proved it true for the case of two particles; it is therefore true for three, four, or any other number of particles, and therefore for any material system.

The kinetic energy of a system referred to its centre of mass is less than its kinetic energy when referred to any other point.

For the latter quantity exceeds the former by a quantity equal to the kinetic energy of a mass equal to that of the whole system moving with the velocity of the centre of mass relative to the other point, and since all kinetic energy is essentially positive, this excess must be positive.

81. Available Kinetic Energy

We have already seen in Article 64 that the mutual action between the parts of a material system cannot change the velocity of the centre of mass of the system. Hence that part of the kinetic energy of the system which depends on the motion of the centre of mass cannot be affected by any action internal to the system. It is therefore impossible, by means of the mutual action of the parts of the system, to convert this part of the energy into work. As far as the system itself is concerned, this energy is unavailable. It can be converted into work only by means of the action between this system and some other material system external to it.

Hence if we consider a material system unconnected with any other system, its available kinetic energy is that which is due to the motions of the parts of the system relative to its centre of mass.

Let us suppose that the action between the parts of the system is such that after a certain time the configuration of the system becomes invariable, and let us call this process the solidification of the system. We have shown that the angular momentum of the whole system is not changed by any mutual action of its parts. Hence if the original angular momentum is zero, the system, when its form becomes invariable, will not rotate about its centre of mass, but if it moves at all will move parallel to itself, and the parts will be at rest relative to the centre of mass. In this case therefore the whole available energy will be converted into work by the mutual action of the parts during the solidification of the system.

If the system has angular momentum, it will have the same angular momentum when solidified. It will therefore rotate about its centre of mass, and will therefore still have energy of motion relative to its centre of mass, and this remaining kinetic energy has not been converted into work.

But if the parts of the system are allowed to separate from one another in directions perpendicular to the axis of the angular momentum of the system, and if the system when thus expanded is solidified, the remaining kinetic energy of rotation round the centre of mass will be less and less the greater the expansion of the system, so that by sufficiently expanding the system [before it is solidified] we may make the remaining kinetic energy as small as we please, so that the whole kinetic energy relative to the centre of mass of the system may be converted into work within the system.

82. Potential Energy

The potential energy of a material system is the capacity which it has of doing work [on other systems] depending on other circumstances than the motion of the system. In other words, potential energy is that energy which is not kinetic.

In the theoretical material system which we build up in our imagination from the fundamental ideas of matter and motion, there are no other conditions present except the configuration and motion of the different masses of which the system is composed. Hence in such a system the circumstances upon which the energy must depend are motion and configuration only, so that, as the kinetic energy depends on the motion, the potential energy must depend on the configuration.

In many real material systems we know that part of the energy does depend on the configuration. Thus the mainspring of a watch has more energy when coiled up than when partially uncoiled, and two bar magnets have more energy when placed side by side with their similar poles turned the same way than when their dissimilar poles are placed next each other.

83. Elasticity

In the case of the spring we may trace the connexion between the coiling of the spring and the force which it exerts somewhat further by conceiving the spring divided (in imagination) into very small parts or elements. When the spring is coiled up, the form of each of these small parts is altered, and such an alteration of the form of a solid body is called a Strain.

In solid bodies strain is accompanied with internal force or stress; those bodies in which the stress depends simply on the strain are called Elastic, and the property of exerting stress when strained is called Elasticity.

We thus find that the coiling of the spring involves the strain of its elements, and that the external force which the spring exerts is the resultant of the stresses in its elements.

We thus substitute for the immediate relation between the coiling of the spring and the force which it exerts, a relation between the strains and stresses of the elements of the spring; that is to say, for a single displacement and a single force, the relation between which may in some cases be of an exceedingly complicated nature, we substitute a multitude of strains and an equal number of stresses, each strain being connected with its corresponding stress by a much more simple relation.

But when all is done, the nature of the connexion between configuration and force remains as mysterious as ever. We can only admit the fact, and if we call all such phenomena phenomena of elasticity, we may find it very convenient to classify them in this way, provided we remember that by the use of the word elasticity we do not profess to explain the cause of the connexion between configuration and energy.

84. Action at a Distance

In the case of the two magnets there is no visible substance connecting the bodies between which the stress exists. The space between the magnets may be filled with air or with water, or we may place the magnets in a vessel and remove the air by an air-pump, till the magnets are left in what is commonly called a vacuum, and yet the mutual action of the magnets will not be altered. We may even place a solid plate of glass or metal or wood between the magnets, and still we find that their mutual action depends simply on their relative position, and is not perceptibly modified by placing any substance between them, unless that substance is one of the magnetic metals. Hence the action between the magnets is commonly spoken of as *action at a distance*.

Attempts have been made, with a certain amount of success[1], to analyse this action at a distance into a continuous distribution of stress in an invisible medium, and thus to establish an analogy between the magnetic action and the action of a spring or a rope in transmitting force; but still the general fact that strains or changes of configuration are accompanied by stresses or internal forces, and that thereby energy is stored up in the system so strained, remains an ultimate fact which has not yet been explained as the result of any more fundamental principle.

85. Theory of Potential Energy more complicated than that of Kinetic Energy

Admitting that the energy of a material system may depend on its configuration, the mode in which it so depends may be much more complicated than the mode in which the kinetic energy depends on the motion of the system. For the kinetic energy may be calculated from the motion of the parts of the system by an invariable method. We multiply the mass of each part by half the square of its velocity, and take the sum of all such products. But the potential energy arising from the mutual action of two parts of the system may depend on the relative position of the parts in a manner which may be different in different instances. Thus when two billiard balls approach each other from a distance, there is no sensible action between them till they come so near one another that certain parts appear to be in contact. To bring the centres of the two balls nearer, the parts in contact must be made to yield, and this requires the expenditure of work.

[1] See Clerk Maxwell's *Treatise on Electricity and Magnetism*, Vol. II, Art. 641. [Modern scrutiny requires a distribution of momentum in the medium, which reveals itself for example in the pressure of radiation, in addition to the stress: cf. appendix to J. H. Poynting's *Collected Papers*. It in fact develops into the guiding tensor principle in the theory of gravitational relativity.]

Hence in this case the potential energy is constant for all distances greater than the distance of first contact, and then rapidly increases when the distance is diminished.

The force between magnets varies with the distance in a very different manner, and in fact we find that it is only by experiment that we can ascertain the form of the relation between the configuration of a system and its potential energy.

86. Application of the Method of Energy to the Calculation of Forces

A complete knowledge of the mode in which the energy of a material system varies when the configuration and motion of the system are made to vary is mathematically equivalent to a knowledge of all the dynamical properties of the system. The mathematical methods by which all the forces and stresses in a moving system are deduced from the single mathematical formula which expresses the energy as a function of the variables have been developed by Lagrange, Hamilton, and other eminent mathematicians, but it would be difficult even to describe them in terms of the elementary ideas to which we restrict ourselves in this book. An outline of these methods is given in my treatise on *Electricity*, Part IV, Chapter V, Article 533*, and the application of these dynamical methods to electromagnetic phenomena is given in the chapters immediately following.

But if we consider only the case of a system at rest it is easy to see how we can ascertain the forces of the system when we know how its energy depends on its configuration.

For let us suppose that an agent external to the system produces a displacement from one configuration to another, then if in the new configuration the system

* Reprinted *infra*, p. 123.

possesses more energy than it did at first, it can have received this increase of energy only from the external agent. This agent must therefore have done an amount of work equal to the increase of energy. It must therefore have exerted force in the direction of the displacement, and the mean value of this force, multiplied into the displacement, must be equal to the work done. Hence the mean value of the force may be found by dividing the increase of energy by the displacement.

If the displacement is large this force may vary considerably during the displacement, so that it may be difficult to calculate its mean value; but since the force depends on the configuration, if we make the displacement smaller and smaller the variation of the force will become smaller and smaller, so that at last the force may be regarded as sensibly constant during the displacement.

If, therefore, we calculate for a given configuration the *rate* at which the energy increases with the displacement, by a method similar to that described in Articles 27, 28, and 33, this rate will be numerically equal to the force exerted by the external agent in the direction of the displacement.

If the energy diminishes instead of increasing as the displacement increases, the system must do work on the external agent, and the force exerted by the external agent must be in the direction opposite to that of displacement.

87. Specification of the [Mode of Action] of Forces

In treatises on dynamics the forces spoken of are usually those exerted by the external agent on the material system. In treatises on electricity, on the other hand, the forces spoken of are usually those exerted by the electrified system against an external agent which prevents the system from moving. It is necessary, therefore, in reading any statement about forces, to ascertain whether the force spoken of is to be regarded from the one point of view or the other.

We may in general avoid any ambiguity by viewing the phenomenon as a whole, and speaking of it as a stress exerted between two points or bodies, and distinguishing it as a tension or a pressure, an attraction or a repulsion, according to its direction. See Article 55.

88. Application to a System in Motion

It thus appears that from a knowledge of the potential energy of a system in every possible configuration we may deduce all the external forces which are required to keep the system in [any given] configuration. If the system is at rest, and if these external forces are the actual forces, the system will remain in equilibrium. If the system is in motion the force acting on each particle is that arising from the connexions of the system (equal and opposite to the external force just calculated), together with any external force which may be applied to it. Hence a complete knowledge of the mode in which the potential energy varies with the configuration would enable us to predict every possible motion of the system under the action of given external forces, provided we were able to overcome the purely mathematical difficulties of the calculation.

89. Application of the Method of Energy to the Investigation of Real Bodies

When we pass from abstract dynamics to physics—from material systems, whose only properties are those expressed by their definitions, to real bodies, whose properties we have to investigate—we find that there are many phenomena which we are not able to explain as changes in the configuration and motion of a material system.

Of course if we begin by assuming that the real bodies are systems composed of matter which agrees in all respects with the definitions we have laid down, we may go on to assert that all phenomena are changes of configuration and motion, though we are not prepared to define the kind of configuration and motion by which the particular phenomena are to be explained. But in accurate science such asserted explanations must be estimated, not by their promises, but by their performances. The configuration and motion of a system are facts capable of being described in an accurate manner, and therefore, in order that the explanation of a phenomenon by the configuration and motion of a material system may be admitted as an addition to our scientific knowledge, the configurations, motions, and forces must be specified, and shown to be consistent with known facts, as well as capable of accounting for the phenomenon.

90. Variables on which the Energy depends

But even when the phenomena we are studying have not yet been explained dynamically, we are still able to make great use of the principle of the conservation of energy as a guide to our researches.

To apply this principle, we in the first place assume that the quantity of energy in a material system depends on the state of that system, so that for a given state there is a definite amount of energy.

Hence the first step is to define the different states of the system, and when we have to deal with real bodies we must define their state with respect not only to the configuration and motion of their visible parts, but if we have reason to suspect that the configuration and motion of their invisible particles influence the visible phenomenon, we must devise some method of estimating the energy thence arising.

Thus pressure, temperature, electric potential, and chemical composition are variable quantities, the values of which serve to specify the state of a body, and in general the energy of the body depends on the values of these and other variables.

91. Energy in Terms of the Variables

The next step in our investigation is to determine how much work must be done by external agency on the body in order to make it pass from one specified state to another.

For this purpose it is sufficient to know the work required to make the body pass from a particular state, which we may call the *standard state*, into any other specified state. The energy in the latter state is equal to that in the standard state, together with the work required to bring it from the standard state into the specified state. The fact that this work is the same through whatever series of states the system has passed from the standard state to the specified state is the foundation of the whole theory of energy.

Since all the phenomena depend on the variations of the energy of the body, and not on its total value, it is unnecessary, even if it were possible, to form any estimate of the energy of the body in its standard state.

92. Theory of Heat

One of the most important applications of the principle of the conservation of energy is to the investigation of the nature of heat.

At one time it was supposed that the difference between the states of a body when hot and when cold was due to the presence of a substance called caloric, which existed in greater abundance in the body when hot than when cold. But the experiments of Rumford on the heat produced by the friction of metal, and of Davy on the melting of ice by friction, have shown that when work is spent in overcoming friction, the amount of heat produced is proportional to the work spent.

The experiments of Hirn have also shown that when heat is made to do work in a steam-engine, part of the heat disappears, and that the heat which disappears is proportional to the work done.

A very careful measurement of the work spent in friction, and of the heat produced, has been made by Joule, who finds that the heat required to raise one pound of water from 39° F. to 40° F. is equivalent to 772 foot-pounds of work at Manchester, or 24,858 foot-poundals.

From this we may find that the heat required to raise one gramme of water from 3° *C*. to 4° C. is 42,000,000 ergs.

93. Heat a Form of Energy

Now, since heat can be produced it cannot be a substance; and since whenever mechanical energy is lost by friction there is a production of heat, and whenever there is a gain of mechanical energy in an engine there is a loss of heat; and since the quantity of energy lost or gained is proportional to the quantity of heat gained or lost, we conclude that heat is a form of energy.

We have also reasons for believing that the minute particles of a hot body are in a state of rapid agitation, that is to say, that each particle is always moving very swiftly, but that the direction of its motion alters so often that it makes little or no progress from one region to another.

If this be the case, a part, and it may be a very large part, of the energy of a hot body must be in the form of kinetic energy.

But for our present purpose it is unnecessary to ascertain in what form energy exists in a hot body; the most important fact is that energy may be measured in the form of heat, and since every kind of energy may be converted into heat, this gives us one of the most convenient methods of measuring it.

94. Energy measured as Heat

Thus when certain substances are placed in contact chemical actions take place, the substances combine in a new way, and the new group of substances has different chemical properties from the original group of substances. During this process mechanical work may be done by the expansion of the mixture, as when gunpowder is fired; an electric current may be produced, as in the voltaic battery; and heat may be generated, as in most chemical actions.

The energy given out in the form of mechanical work may be measured directly, or it may be transformed into heat by friction. The energy spent in producing the electric current may be estimated as heat by causing the current to flow through a conductor of such a form that the heat generated in it can easily be measured. Care must be taken that no energy is transmitted to a distance in the form of sound or radiant heat without being duly accounted for.

The energy remaining in the mixture, together with the energy which has escaped, must be equal to the original energy.

Andrews, Favre and Silbermann, [Julius Thomsen,] and others, have measured the quantity of heat produced when a certain quantity of oxygen or of chlorine combines with its equivalent of other substances. These measurements enable us to calculate the excess of the energy which the substances concerned had in their original state, when uncombined, above that which they have after combination.

95. Scientific Work to be done

Though a great deal of excellent work of this kind has already been done, the extent of the field hitherto investigated appears quite insignificant when we consider the boundless variety and complexity of the natural bodies with which we have to deal.

In fact the special work which lies before the physical inquirer in the present state of science is the determination of the quantity of energy which enters or leaves a material system during the passage of the system from its standard state to any other definite state.

96. History of the Doctrine of Energy

The scientific importance of giving a name to the quantity which we call kinetic energy seems to have been first recognised by Leibniz, who gave to the product of the mass by the square of the velocity the name of *Vis Viva*. This is twice the kinetic energy.

Newton, in the "Scholium to the Laws of Motion," expresses the relation between the rate at which work is done by the external agent, and the rate at which it is given out, stored up, or transformed by any machine or other material system, in the following statement, which he makes in order to show the wide extent of the application of the Third Law of Motion.

"If the action of the external agent is estimated by the product of its force into its velocity, and the reaction of the resistance in the same way by the product of the velocity of each part of the system into the resisting force arising from friction, cohesion, weight, and acceleration, the action and reaction will be equal to each other, whatever be the nature and motion of the system." That this statement of Newton's implicitly contains nearly the whole doctrine of energy was first pointed out by Thomson and Tait*.

The words Action and Reaction as they occur in the enunciation of the Third Law of Motion are explained to mean Forces, that is to say, they are the opposite aspects of one and the same Stress.

In the passage quoted above a new and different sense is given to these words by estimating Action and Reaction by the product of a force into the velocity of

* *Treatise on Natural Philosophy*, vol. I, 1867, § 268.

"Newton, in a Scholium to his Third Law of Motion, has stated the relation between work and kinetic energy in a manner so perfect that it cannot be improved, but at the same time with so little apparent effort or desire to attract attention that no one seems to have been struck with the great importance of the passage till it was pointed out recently (1867) by Thomson and Tait." Clerk Maxwell's *Theory of Heat*, ch. IV on "Elementary Dynamical Principles," p. 91.

its point of application. According to this definition the Action of the external agent is the rate at which it does work. This is what is meant by the Power of a steam-engine or other prime mover. It is generally expressed by the estimated number of ideal horses which would be required to do the work at the same rate as the engine, and this is called the Horse-power of the engine.

When we wish to express by a single word the rate at which work is done by an agent we shall call it the Power of the agent, defining the power as the work done in the unit of time.

The use of the term Energy, in a precise and scientific sense, to express the quantity of work which a material system can do, was introduced by Dr Young*.

97. On the Different Forms of Energy

The energy which a body has in virtue of its motion is called kinetic energy.

A system may also have energy in virtue of its configuration, if the forces of the system are such that the system will do work against external resistance while it passes into another configuration. This energy is called Potential Energy. Thus when a stone has been lifted to a certain height above the earth's surface, the system of two bodies, the stone and the earth, has potential energy, and is able to do a certain amount of work during the descent of the stone. This potential energy is due to the fact that the stone and the earth attract each other, so that work has to be spent by the man who lifts the stone and draws it away from the earth, and after the stone is lifted the attraction between the earth and the stone is capable of doing work as the stone descends. This kind of energy, therefore, depends upon the work which the forces of the system would do

* *Lectures on Natural Philosophy* [1807], Lecture VIII.

if the parts of the system were to yield to the action of these forces. This is called the "Sum of the Tensions" by Helmholtz in his celebrated memoir on the "Conservation of Energy."* Thomson called it Statical Energy; it has also been called Energy of Position; but Rankine introduced the term Potential Energy†—a very felicitous expression, since it not only signifies the energy which the system has not in actual possession, but only has the power to acquire, but it also indicates its connexion with what has been called (on other grounds) the Potential Function‡.

The different forms in which energy has been found to exist in material systems have been placed in one or other of these two classes—Kinetic Energy, due to motion, and Potential Energy, due to configuration.

Thus a hot body, by giving out heat to a colder body, may be made to do work by causing the cold body to expand in opposition to pressure. A material system, therefore, in which there is a non-uniform distribution of temperature has the capacity of doing work, or energy. This energy is now believed to be kinetic energy, due to a motion of agitation in the smallest parts of the hot body.

Gunpowder has energy, for when fired it is capable of setting a cannon-ball in motion. The energy of gunpowder is Chemical Energy, arising from the power which the constituents of gunpowder possess of arranging themselves in a new manner when exploded, so as to occupy a much larger volume than the gunpowder does. In the present state of science chemists figure to themselves chemical action as a rearrangement of particles under the action of forces tending to produce

* Berlin, 1847: translated in Taylor's *Scientific Memoirs*, Feb. 1853. [Remarkable mainly for its wide ramifications into electric and chemical theory.]

† The *vis potentialis* of Daniel Bernoulli, as contrasted with *vis viva*, e.g. for the case of a bent spring; cf. Euler, *De Curvis Elasticis*, in Appendix to *Solutio Problematis Isoperimetrici*... (1744).

‡ The term Potential was employed independently by Gauss and by Green, and so probably originated with D. Bernoulli.

this change of arrangement. From this point of view, therefore, chemical energy is potential energy.

Air, compressed in the chamber of an air-gun, is capable of propelling a bullet. The energy of compressed air was at one time supposed to arise from the mutual repulsion of its particles. If this explanation were the true one its energy would be potential energy. In more recent times it has been thought that the particles of the air are in a state of motion, and that its pressure is caused by the impact of these particles on the sides of the vessel. According to this theory the energy of compressed air is kinetic energy.

There are thus many different modes in which a material system may possess energy, and it may be doubtful in some cases whether the energy is of the kinetic or the potential form. The nature of energy, however, is the same in whatever form it may be found. The quantity of energy can always be expressed as equated to that of a body of a definite mass moving with a definite velocity.

CHAPTER VI

RECAPITULATION

98. Retrospect of Abstract Dynamics

We have now gone through that part of the fundamental science of the motion of matter which we have been able to treat in a manner sufficiently elementary to be consistent with the plan of this book.

It remains for us to take a general view of the relations between the parts of this science, and of the whole to other physical sciences, and this we can now do in a more satisfactory way than we could before we had entered into the subject.

99. Kinematics

We began with kinematics, or the science of pure motion. In this division of the subject the ideas brought before us are those of space and time. The only attribute of matter which comes before us is its continuity of existence in space and time—the fact, namely, that every particle of matter, at any instant of time, is in one place and in one only, and that its change of place during any interval of time is accomplished by moving along a continuous path.

Neither the force which affects the motion of the body, nor the mass of the body, on which the amount of force required to produce the motion depends, come under our notice in the pure science of motion.

100. Force

In the next division of the subject force is considered in the aspect of that which alters the motion of a mass.

If we confine our attention to a single body, our investigation enables us, from observation of its motion, to determine the direction and magnitude of the resultant force which acts on it, and this investigation is the exemplar and type of all researches undertaken for the purpose of the discovery and measurement of physical forces.

But this may be regarded as a mere application of the definition of a force, and not as a new physical truth.

It is when we come to define equal forces as those which produce equal rates of acceleration in the same mass, and equal masses as those which are equally accelerated by equal forces, that we find that these definitions of equality amount to the assertion of the physical truth, that the comparison of quantities of matter by the forces required to produce in them a given acceleration is a method which always leads to consistent results, whatever be the absolute values of the forces and the accelerations.

101. Stress

The next step in the science of force is that in which we pass from the consideration of a force as acting on a body, to that of its being one aspect of that mutual action between two bodies, which is called by Newton Action and Reaction, and which is now more briefly expressed by the single word Stress.

102. Relativity of Dynamical Knowledge

Our whole progress up to this point may be described as a gradual development of the doctrine of relativity of all physical phenomena. Position we must evidently acknowledge to be relative, for we cannot describe the position of a body in any terms which do not express relation. The ordinary language about motion and rest does not so completely exclude the notion of their being measured absolutely, but the reason of this is, that in our ordinary language we tacitly assume that the earth is at rest.

As our ideas of space and motion become clearer, we come to see how the whole body of dynamical doctrine hangs together in one consistent system.

Our primitive notion may have been that to know absolutely where we are, and in what direction we are going, are essential elements of our knowledge as conscious beings.

But this notion, though undoubtedly held by many wise men in ancient times, has been gradually dispelled from the minds of students of physics.

There are no landmarks in space; one portion of space is exactly like every other portion, so that we cannot tell where we are. We are, as it were, on an unruffled sea, without stars, compass, soundings, wind, or tide, and we cannot tell in what direction we are going. We have no log which we can cast out to take a dead reckoning by; we may compute our rate of motion with respect to the neighbouring bodies, but we do not know how these bodies may be moving in space.

103. Relativity of Force

We cannot even tell what force may be acting on us; we can only tell the difference between the force acting on one thing and that acting on another*.

We have an actual example of this in our every-day experience. The earth moves round the sun in a year at a distance of 91,520,000 miles or $1{\cdot}473 \times 10^{13}$ centimetres†. It follows from this that a force is exerted on the earth in the direction of the sun, which produces an acceleration of the earth in the direction of the sun of about 0·019 in feet and seconds, or about $\frac{1}{1680}$ of the intensity of gravity at the earth's surface.

A force equal to the sixteen-hundredth part of the weight of a body might be easily measured by known experimental methods, especially if the direction of this

* See Appendix I; especially p. 143.

† More modern values are $9{\cdot}28 \times 10^7$ miles, or $1{\cdot}494 \times 10^{13}$ cm.

force were differently inclined to the vertical at different hours of the day.

Now, if the attraction of the sun were exerted upon the solid part of the earth, as distinguished from the movable bodies on which we experiment, a body suspended by a string, and moving with the earth, would indicate the difference between the solar action on the body, and that on the earth as a whole.

If, for example, the sun attracted the earth and not the suspended body, then at sunrise the point of suspension, which is rigidly connected with the earth, would be drawn towards the sun, while the suspended body would be acted on only by the earth's attraction, and the string would appear to be deflected away from the sun by a sixteen-hundredth part of the length of the string. At sunset the string would be deflected away from the setting sun by an equal amount; and as the sun sets at a different point of the compass from that at which he rises the deflexions of the string would be in different directions, and the difference in the position of the plumb-line at sunrise and sunset would be easily observed.

But instead of this, the attraction of gravitation is exerted upon all kinds of matter equally at the same distance from the attracting body. At sunrise and sunset the centre of the earth and the suspended body are nearly at the same distance from the sun, and no deflexion of the plumb-line due to the sun's attraction can be observed at these times. The attraction of the sun, therefore, in so far as it is exerted equally upon all bodies on the earth, produces no effect on their relative motions. It is only the differences of the intensity and direction of the attraction acting on different parts of the earth which can produce any effect, and these differences are so small for bodies at moderate distances that it is only when the body acted on is very large, as in the case of the ocean, that their effect becomes perceptible in the form of tides.

104. Rotation

In what we have hitherto said about the motion of bodies, we have tacitly assumed that in comparing one configuration of the system with another, we are able to draw a line in the final configuration parallel to a line in the original configuration. In other words, we assume that there are certain directions in space which may be regarded as constant, and to which other directions may be referred during the motion of the system.

In astronomy, a line drawn from the earth to a star may be considered as fixed in direction, because the relative motion of the earth and the star is in general so small compared with the distance between them that the change of direction, even in a century, is very small. But it is manifest that all such directions of reference must be indicated by the configuration of a material system existing in space, and that if this system were altogether removed, the original directions of reference could never be recovered.

But though it is impossible to determine the absolute velocity of a body in space, it is possible to determine whether the direction of a line in a material system is constant or variable.

For instance, it is possible by observations made on the earth alone, without reference to the heavenly bodies, to determine whether the earth is rotating or not.

So far as regards the geometrical configuration of the earth and the heavenly bodies, it is evidently all the same*

> "Whether the sun, predominant in heaven,
> Rise on the earth, or earth rise on the sun;
> He from the east his flaming road begin,
> Or she from west her silent course advance
> With inoffensive pace that spinning sleeps
> On her soft axle, while she paces even,
> And bears thee soft with the smooth air along."

* From the discussion on the celestial motions in *Paradise Lost* (Book VIII, lines 160–6): Milton's interview with Galileo when as a young man he visited Italy may be recalled.

The distances between the bodies composing the universe, whether celestial or terrestrial, and the angles between the lines joining them, are all that can be ascertained without an appeal to dynamical principles, and these will not be affected if any motion of rotation of the whole system, similar to that of a rigid body about an axis, is combined with the actual motion; so that from a geometrical point of view the Copernican system, according to which the earth rotates, has no advantage, except that of simplicity, over that in which the earth is supposed to be at rest, and the apparent motions of the heavenly bodies to be their absolute motions.

Even if we go a step further, and consider the dynamical theory of the earth rotating round its axis, we may account for its oblate figure, and for the equilibrium of the ocean and of all other bodies on its surface on either of two hypotheses—that of the motion of the earth round its axis, or that of the earth not rotating, but caused to assume its oblate figure by a force acting outwards in all directions from its axis, the intensity of this force increasing as the distance from the axis increases. Such a force, if it acted on all kinds of matter alike, would account not only for the oblateness of the earth's figure, but for the conditions of equilibrium of all bodies at rest with respect to the earth.

It is only when we go further still, and consider the phenomena of bodies which are in motion with respect to the earth*, that we are really constrained to admit that the earth rotates.

105. Newton's Determination of the Absolute Velocity of Rotation

Newton was the first to point out that the absolute motion of rotation of the earth might be demonstrated by experiments on the rotation of a material system.

* As in Art. 105. See also Appendix I, p. 142.

For instance, if a bucket of water is suspended from a beam by a string, and the string twisted so as to keep the bucket spinning round a vertical axis, the water will soon spin round at the same rate as the bucket, so that the system of the water and the bucket turns round its axis like a solid body.

The water in the spinning bucket rises up at the sides, and is depressed in the middle, showing that in order to make it move in a circle a pressure must be exerted towards the axis. This concavity of the surface depends on the absolute motion of rotation of the water and not on its relative rotation.

For instance, it does not depend on the rotation relative to the bucket. For at the beginning of the experiment, when we set the bucket spinning, and before the water has taken up the motion, the water and the bucket are in relative motion, but the surface of the water is flat, because the water is not rotating, but only the bucket.

When the water and the bucket rotate together, there is no motion of the one relative to the other, but the surface of the water is hollow, because it is rotating.

When the bucket is stopped, as long as the water continues to rotate its surface remains hollow, showing that it is still rotating though the bucket is not.

It is manifestly the same, as regards this experiment, whether the rotation be in the direction of the hands of a watch or the opposite direction, provided the rate of rotation is the same.

Now let us suppose this experiment tried at the North Pole. Let the bucket be made, by a proper arrangement of clockwork, to rotate either in the direction of the hands of a watch, or in the opposite direction, at a perfectly regular rate.

If it is made to turn round by clockwork once in twenty-four hours (sidereal time) the way of the hands of a watch laid face upwards, it will be rotating as regards the earth, but not rotating as regards the stars.

If the clockwork is stopped, it will rotate with respect to the stars, but not with respect to the earth.

Finally, if it is made to turn round once in twenty-four hours (sidereal time) in the opposite direction, it will be rotating with respect to the earth at the same rate as at first, but instead of being free from rotation as respects the stars, it will be rotating at the rate of one turn in twelve hours.

Hence if the earth is at rest, and the stars moving round it, the form of the surface will be the same in the first and last case; but if the earth is rotating, the water will be rotating in the last case but not in the first, and this will be made manifest by the water rising higher at the sides in the last case than in the first.

The surface of the water will not be really concave in any of the cases supposed, for the effect of gravity acting towards the centre of the earth is to make the surface convex, as the surface of the sea is, and the rate of rotation in our experiment is not sufficiently rapid to make the surface concave. It will only make it slightly less convex than the surface of the sea in the last case, and slightly more convex in the first.

But the difference in the form of the surface of the water would be so exceedingly small, that with our methods of measurement it would be hopeless to attempt to determine the rotation of the earth in this way.

106. Foucault's Pendulum

The most satisfactory method of making an experiment for this purpose is that devised by M. Foucault*.

A heavy ball is hung from a fixed point by a wire, so that it is capable of swinging like a pendulum in any vertical plane passing through the fixed point.

* Nowadays the fixity of direction in space of the plane of rotation of a rapidly spinning wheel, freely pivoted, a method also originated by Foucault, would reveal it most readily. Cf. Art. 71. The gyrostatic compass interacts with the earth's rotation, on the same principle.

In starting the pendulum care must be taken that the wire, when at the lowest point of the swing, passes exactly through the position it assumes when hanging vertically. If it passes on one side of this position, it will return on the other side, and this motion of the pendulum round the vertical instead of through the vertical must be carefully avoided, because we wish to get rid of all motions of rotation either in one direction or the other.

Let us consider the angular momentum of the pendulum about the vertical line through the fixed point.

At the instant at which the wire of the pendulum passes through the vertical line, the angular momentum about the vertical line is zero.

The force of gravity always acts parallel to this vertical line, so that it cannot produce angular momentum round it. The tension of the wire always acts through the fixed point, so that it cannot produce angular momentum about the vertical line.

Hence the pendulum can never acquire angular momentum about the vertical line through the point of suspension.

Hence when the wire is out of the vertical, the vertical plane through the centre of the ball and the point of suspension cannot be rotating; for if it were, the pendulum would have an angular momentum about the vertical line*.

Now let us suppose this experiment performed at the North Pole. The plane of vibration of the pendulum will remain absolutely constant in direction, so that if the earth rotates, the rotation of the earth will be made manifest.

* But if from want of precaution the ball described an open elliptic curve, however elongated, this curve of vibration would rotate spontaneously, through an angle $\frac{3}{4}\Omega$ in each revolution of the ball, and in the same direction, where Ω is the (small) extent of the conical angle traced out by the wire. This may readily mask the effect of the earth's rotation. If the bob were free to revolve on the wire as axis, that body would turn through Ω in each revolution.

We have only to draw a line on the earth parallel to the plane of vibration, and to compare the position of this line with that of the plane of vibration at a subsequent time.

As a pendulum of this kind properly suspended will swing for several hours, it is easy to ascertain whether the position of the plane of vibration is constant as regards the earth, as it would be if the earth is at rest, or constant as regards the stars, if the stars do not move round the earth.

We have supposed, for the sake of simplicity in the description, that the experiment is made at the North Pole. It is not necessary to go there in order to demonstrate the rotation of the earth. The only region where the experiment will not show it is at the equator.

At every other place the pendulum will indicate the rate of rotation of the earth with respect to the vertical line at that place. If at any instant the plane of the pendulum passes through a star near the horizon either rising or setting, it will continue to pass through that star as long as it is near the horizon. That is to say, the horizontal part of the apparent motion of a star on the horizon is equal to the rate of rotation of the plane of vibration of the pendulum.

It has been observed that the plane of vibration appears to rotate in the opposite direction in the southern hemisphere, and by a comparison of the rates at various places the actual time of rotation of the earth has been deduced without reference to astronomical observations. The mean value, as deduced from these experiments by Messrs Galbraith and Haughton in their *Manual of Astronomy*, is 23 hours 53 minutes 37 seconds. The true time of rotation of the earth is 23 hours 56 minutes 4 seconds mean solar time.

107. MATTER AND ENERGY*

All that we know about matter relates to the series of phenomena in which energy is transferred from one portion of matter to another, till in some part of the series our bodies are affected, and we become conscious of a sensation.

By the mental process which is founded on such sensations we come to learn the conditions of these sensations, and to trace them to objects which are not part of ourselves, but in every case the fact that we learn is the mutual action between bodies. This mutual action we have endeavoured to describe in this treatise. Under various aspects it is called Force, Action and Reaction, and Stress, and the evidence of it is the change of the motion of the bodies between which it acts.

The process by which stress produces change of motion is called Work, and, as we have already shown, work may be considered as the transference of Energy from one body or system to another.

Hence, as we have said, we are acquainted with matter only as that which may have energy communicated to it from other matter, and which may, in its turn, communicate energy to other matter.

Energy, on the other hand, we know only as that which in all natural phenomena is continually passing from one portion of matter to another.

108. TEST OF A MATERIAL SUBSTANCE

Energy cannot exist except in connexion with matter. Hence since, in the space between the sun and the earth, the luminous and thermal radiations, which have left the sun and which have not reached the earth, possess energy, the amount of which per cubic mile can be measured, this energy must belong to matter existing

* See Appendix II.

in the interplanetary spaces, and since it is only by the light which reaches us that we become aware of the existence of the most remote stars, we conclude that the matter which transmits light is disseminated through the whole of the visible universe.

109. Energy not capable of Identification

We cannot identify a particular portion of energy, or trace it through its transformations. It has no individual existence, such as that which we attribute to particular portions of matter.

The transactions of the material universe appear to be conducted, as it were, on a system of credit*. Each transaction consists of the transfer of so much credit or energy from one body to another. This act of transfer or payment is called work. The energy so transferred does not retain any character by which it can be identified when it passes from one form to another.

110. Absolute Value of the Energy of a Body unknown

The energy of a material system can only be estimated in a relative manner.

In the first place, though the energy of the motion of the parts relative to the centre of mass of the system may be accurately defined, the whole energy consists of this together with the energy of a mass equal to that of the whole system moving with the velocity of the centre of mass. Now this latter velocity—that of the centre of mass—can be estimated only with reference to some body external to the system, and the value which we assign to this velocity will be different according to the body which we select as our origin.

Hence the estimated kinetic energy of a material

* Except perhaps that credit can be artificially increased, or inflated.

system contains a part, the value of which cannot be determined except by the arbitrary selection of an origin. The only origin which would not be arbitrary is the centre of mass of the material universe, but this is a point the position and motion of which are quite unknown to us.

111. LATENT ENERGY

But the energy of a material system is indeterminate for another reason. We cannot reduce the system to a state in which it has no energy, and any energy which is never removed from the system must remain unperceived by us, for it is only as it enters or leaves the system that we can take any account of it.

We must, therefore, regard the energy of a material system as a quantity of which we may ascertain the increase or diminution as the system passes from one definite condition to another. The absolute value of the energy in the standard condition is unknown to us, and it would be of no value to us if we did know it, as all phenomena depend on the variations of the energy, and not on its absolute value.

112. A COMPLETE DISCUSSION OF ENERGY WOULD INCLUDE THE WHOLE OF PHYSICAL SCIENCE

The discussion of the various forms of energy—gravitational, electro-magnetic, molecular, thermal, etc. —with the conditions of the transference of energy from one form to another, and the constant dissipation of the energy available for producing work, constitutes the whole of physical science, in so far as it has been developed in the dynamical form under the various designations of Astronomy, Electricity, Magnetism, Optics, Theory of the Physical States of Bodies, Thermo-dynamics, and Chemistry.

CHAPTER VII

THE PENDULUM AND GRAVITY

113. On Uniform Motion in a Circle

Let M (fig. 11) be a body moving in a circle with velocity V.

Let $OM = r$ be the radius of the circle.

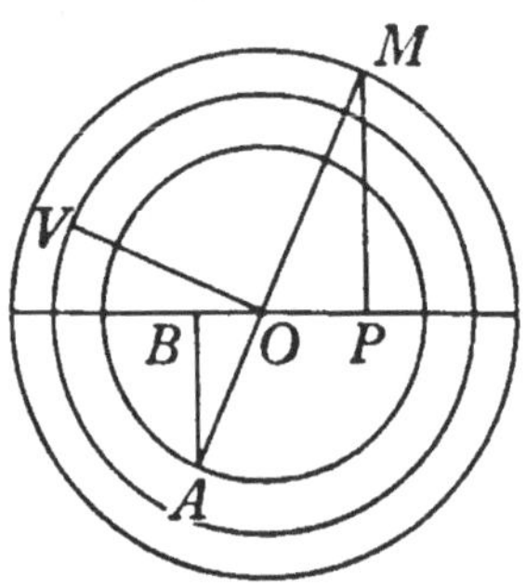

Fig. 11.

The direction of the velocity of M is that of the tangent to the circle. Draw OV parallel to this direction through the centre of the circle and equal to the distance described in unit of time with velocity V, then $OV = V$.

If we take O as the origin of the diagram of velocity, V will represent the velocity of the body at M.

As the body moves round the circle, the point V will also describe a circle, and the velocity of the point V will be to that of M as OV to OM.

If, therefore, we draw OA in MO produced, and therefore parallel to the direction of motion of V, and make OA a third proportional to OM and OV, and if we assume O as the origin of the diagram of rate of acceleration, then the point A will represent the velocity of the point V, or, what is the same thing, the rate of acceleration of the point M.

Hence, when a body moves with uniform velocity in a circle, its acceleration is directed towards the centre of the circle, and is a third proportional to the radius of the circle and the velocity of the body.

The force acting on the body M is equal to the

product of this acceleration into the mass of the body, or if F be this force

$$F = \frac{MV^2}{r}.$$

114. Centrifugal Force

This force F is that which must act on the body M in order to keep it in the circle of radius r, in which it is moving with velocity V.

The direction of this force is towards the centre of the circle.

If this force is applied by means of a string fastened to the body, the string will be in a state of tension. To a person holding the other end of the string this tension will appear to be directed towards the body M, as if the body M had a tendency to move away from the centre of the circle which it is describing.

Hence this latter force is often called Centrifugal Force.

The force which really acts on the body, being directed towards the centre of the circle, is called Centripetal Force, and in some popular treatises the centripetal and centrifugal forces are described as opposing and balancing each other. But they are merely the different aspects of the same stress [acting in the string].

115. Periodic Time

The time of describing the circumference of the circle is called the Periodic Time. If π represents the ratio of the circumference of a circle to its diameter, which is 3·14159.., the circumference of a circle of radius r is $2\pi r$; and since this is described in the periodic time T with velocity V, we have

$$2\pi r = VT.$$

Hence
$$F = 4\pi^2 M \frac{r}{T^2}.$$

The rate of circular motion is often expressed by the number of revolutions in unit of time. Let this number [the frequency] be denoted by n, then

$$nT = 1$$

and
$$F = 4\pi^2 Mrn^2.$$

116. On Simple Harmonic Vibrations

If while the body M (fig. 11) moves in a circle with uniform velocity another point P moves in a fixed diameter of the circle, so as to be always at the foot of the perpendicular from M on that diameter, the body P is said to execute Simple Harmonic Vibrations.

The radius, r, of the circle is called the Amplitude of the vibration.

The periodic time of M is called the Periodic Time of Vibration.

The angle which OM makes with the positive direction of the fixed diameter is called the Phase of the vibration.

117. On the Force acting on the Vibrating Body

The only difference between the motions of M and P is that M has a vertical motion compounded with a horizontal motion which is the same as that of P. Hence the velocity and the acceleration of the two bodies differ only with respect to the vertical part of the velocity and acceleration of M.

The acceleration of P is therefore the horizontal component of that of M, and since the acceleration of M is represented by OA, which is in the direction of MO produced, the acceleration of P will be represented by OB, where B is the foot of the perpendicular from A on the horizontal diameter. Now by similar triangles OMP, OAB

$$OM : OA = OP : OB.$$

But $\quad OM = r$ and $OA = -4\pi^2 \frac{r}{T^2}$.

Hence $\quad OB = -\frac{4\pi^2}{T^2} OP = -4\pi^2 n^2 OP.$

In simple harmonic vibration, therefore, the acceleration is always directed towards the centre of vibration, and is equal to the distance from that centre multiplied by $4\pi^2n^2$, and if the mass of the vibrating body is P, the force acting on it at a distance x from O is $4\pi^2n^2Px$.

It appears, therefore, that a body which executes simple harmonic vibrations in a straight line is acted on by a force which varies as the distance from the centre of vibration, and the value of this force at a given distance depends only on that distance, on the mass of the body, and on the square of the number of vibrations in unit of time, and is independent of the amplitude of the vibrations.

118. Isochronous Vibrations

It follows from this that if a body moves in a straight line and is acted on by a force directed towards a fixed point on the line and varying as the distance from that point, it will execute simple harmonic vibrations, the periodic time of which will be the same whatever the amplitude of vibration.

If for a particular kind of displacement of a body, as turning round an axis, the force tending to bring it back to a given position varies as the displacement, the body will execute simple harmonic vibrations about that position, the periodic time of which will be independent of their amplitude.

Vibrations of this kind, which are executed in the same time whatever be their amplitude, are called Isochronous Vibrations.

119. Potential Energy of the Vibrating Body

The velocity of the body when it passes through the point of equilibrium is equal to that of the body moving in the circle, or

$$V = 2\pi r n,$$

where r is the amplitude of vibration and n is the number of double vibrations per second.

Hence the kinetic energy of the vibrating body at the point of equilibrium is

$$\tfrac{1}{2}MV^2 = 2\pi^2 M r^2 n^2,$$

where M is the mass of the body.

At the extreme elongation, where $x = \mathrm{r}$, the velocity, and therefore the kinetic energy, of the body is zero. The diminution of kinetic energy must correspond to an equal increase of potential energy. Hence if we reckon the potential energy from the configuration in which the body is at its point of equilibrium, its potential energy when at a distance r from this point is $2\pi^2 M n^2 r^2$.

This is the potential energy of a body which vibrates isochronously, and executes n double vibrations per second when it is at rest at the distance, r, from the point of equilibrium. As the potential energy does not depend on the motion of the body, but only on its position, we may write it

$$2\pi^2 M n^2 x^2,$$

where x is the distance from the point of equilibrium.

120. The Simple Pendulum

The simple pendulum consists of a small heavy body called the bob, suspended from a fixed point by a fine string of invariable length. The bob is supposed to be so small that its motion may be treated as that of a material particle, and the string is supposed to be so fine that we may neglect its mass and weight. The bob is set in motion so as to swing through a small angle in a vertical plane. Its path, therefore, is an arc of a circle, whose centre is the point of suspension, O, and whose radius is the length of the string, which we shall denote by l.

Let O (fig. 12) be the point of suspension and OA the position of the pendulum when hanging vertically. When the bob is at M it is higher than when it is at A by the height $AP = \frac{AM^2}{AB}$ where AM is the chord of the arc AM and $AB = 2l$.

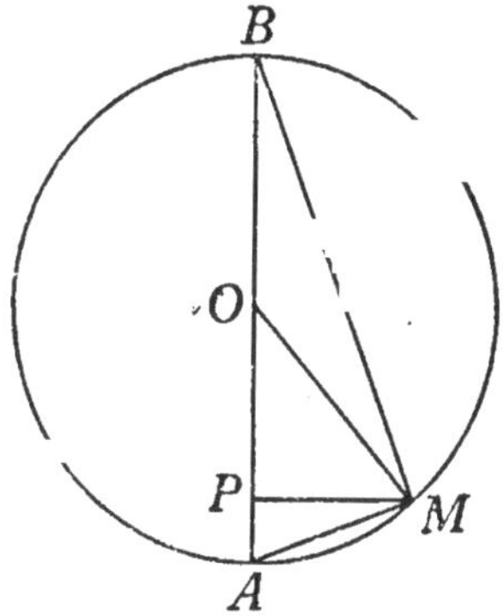

Fig. 12.

If M be the mass of the bob and g the intensity of gravity the weight of the bob will be Mg and the work done against gravity during the motion of the bob from A to M will be $MgAP$. This, therefore, is the potential energy of the pendulum when the bob is at M, reckoning the energy zero when the bob is at A.

We may write this energy

$$\frac{Mg}{2l} AM^2.$$

The potential energy of the bob when displaced through any arc varies as the square of the chord of that arc.

If it had varied as the square of the arc itself in which the bob moves, the vibrations would have been strictly isochronous. As the potential energy varies more slowly than the square of the arc, the period of each vibration will be greater when the amplitude is greater.

For very small vibrations, however, we may neglect the difference between the chord and the arc, and denoting the arc by x we may write the potential energy

$$\frac{Mg}{2l}x^2.$$

But we have already shown that in harmonic vibrations the potential energy is $2\pi^2 Mn^2x^2$.

Equating these two expressions and clearing fractions we find

$$g = 4\pi^2 n^2 l,$$

where g is the intensity of gravity, π is the ratio of the circumference of a circle to its diameter, n is the number of vibrations of the pendulum in unit of time, and l is the length of the pendulum.

121. A Rigid Pendulum

If we could construct a pendulum with a bob so small and a string so fine that it might be regarded for practical purposes as a simple pendulum, it would be easy to determine g by this method. But all real pendulums have bobs of considerable size, and in order to preserve the length invariable the bob must be connected with the point of suspension by a stout rod, the mass of which cannot be neglected. It is always possible, however, to determine the length of a simple pendulum whose vibrations would be executed in the same manner as those of a pendulum of any shape.

The complete discussion of this subject would lead us into calculations beyond the limits of this treatise. We may, however, arrive at the most important result without calculation as follows.

The motion of a rigid body in one plane may be completely defined by stating the motion of its centre of mass, and the motion of the body round its centre of mass.

The force required to produce a given change in the motion of the centre of mass depends only on the mass of the body (Art. 63).

The moment required to produce a given change of angular velocity about the centre of mass depends on the distribution of the mass, being greater the further the different parts of the body are from the centre of mass.

If, therefore, we form a system of two particles rigidly connected, the sum of the masses being equal to the mass of a pendulum, their centre of mass coinciding with that of the pendulum, and their distances from the centre of mass being such that a couple of the same moment is required to produce a given rotatory motion about the centre of mass of the new system as about that of the pendulum, then the new system will for motions in a certain plane be dynamically equivalent to the given pendulum, that is, if the two systems are moved in the same way the forces required to guide the motion will be equal. Since the two particles may have any ratio, provided the sum of their masses is equal to the mass of the pendulum, and since the line joining them may have any direction provided it passes through the centre of mass, we may arrange them so that one of the particles corresponds to any given point of the pendulum, say, the point of suspension P (fig. 13). The mass of this particle and the position and mass of the other at Q will be determinate. The position of the second particle, Q, is called the Centre of Oscillation. Now in the system of two particles, if one of them, P, is fixed and the other, Q, allowed to swing under the action of gravity, we have a simple pendulum. For one of the particles, P, acts as the point of suspension, and the other, Q, is at an invariable distance from it, so that the connexion between them is the same as if they were united by a string of length $l = PQ$.

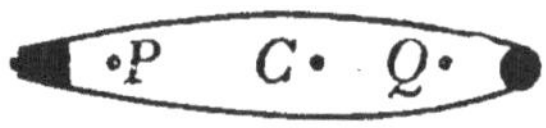

Fig. 13.

Hence a pendulum of any form swings in exactly the same manner as a simple pendulum whose length is the distance from the centre of suspension to the centre of oscillation.

122. Inversion of the Pendulum

Now let us suppose the system of two particles inverted, Q being made the point of suspension and P being made to swing. We have now a simple pendulum of the same length as before. Its vibrations will therefore be executed in the same time. But it is dynamically equivalent to the pendulum suspended by its centre of oscillation.

Hence if a pendulum be inverted and suspended by its centre of oscillation its vibrations will have the same period as before, and the distance between the centre of suspension and that of oscillation will be equal to that of a simple pendulum having the same time of vibration.

It was in this way that Captain Kater determined the length of the simple pendulum which vibrates seconds.

He constructed a pendulum which could be made to vibrate about two knife edges, on opposite sides of the centre of mass and at *unequal* distances from it.

By certain adjustments, he made the time of vibration the same whether the one knife edge or the other were the centre of suspension. The length of the corresponding simple pendulum was then found by measuring the distance between the knife edges.

123. Illustration of Kater's Pendulum

The principle of Kater's Pendulum may be illustrated by a very simple and striking experiment. Take a flat board of any form (fig. 14), and drive a piece of wire through it near its edge, and allow it to hang in a vertical plane, holding the ends of the wire by the finger and thumb. Take a small bullet, fasten it to the end of a thread and allow the thread to pass over the wire, so that the bullet hangs close to the board. Move the hand by which you hold the wire horizontally in the plane of the board, and observe whether the board moves forwards or backwards with respect to the bullet. If it moves forwards lengthen the string, if backwards shorten it till the bullet and the board move together. Now mark the point of the board opposite the centre of the bullet and fasten the string to the wire. You will find that if you hold the wire by the ends and move it in any manner, however sudden and irregular, in the plane of the board, the bullet will never quit the marked spot on the board.

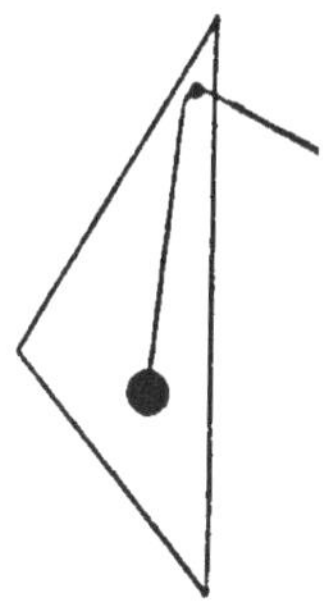

Fig. 14.

Hence this spot is called the centre of oscillation, because when the board is oscillating about the wire when fixed it oscillates as if it consisted of a single particle placed at the spot.

It is also called the centre of percussion, because if the board is at rest and the wire is suddenly moved horizontally the board will at first begin to rotate about the spot as a centre.

124. Determination of the Intensity of Gravity

The most direct method of determining g is, no doubt, to let a body fall and find what velocity it has gained in a second, but it is very difficult to make accurate observations of the motion of bodies when their velocities are so great as 981 centimetres per second, and besides, the experiment would have to be conducted in a vessel from which the air has been exhausted, as the resistance of the air to such rapid motion is very considerable, compared with the weight of the falling body.

The experiment with the pendulum is much more satisfactory. By making the arc of vibration very small, the motion of the bob becomes so slow that the resistance of the air can have very little influence on the time of vibration. In the best experiments the pendulum is swung in an air-tight vessel from which the air is exhausted.

Besides this, the motion repeats itself, and the pendulum swings to and fro hundreds, or even thousands, of times before the various resistances to which it is exposed reduce the amplitude of the vibrations till they can no longer be observed.

Thus the actual observation consists not in watching the beginning and end of one vibration, but in determining the duration of a series of many hundred vibrations, and thence deducing the time of a single vibration.

The observer is relieved from the labour of counting the whole number of vibrations, and the measurement is made one of the most accurate in the whole range of practical science by the following method.

125. METHOD OF OBSERVATION

A pendulum clock is placed behind the experimental pendulum, so that when both pendulums are hanging vertically the bob, or some other part of the experimental pendulum, just hides a white spot on the clock pendulum, as seen by a telescope fixed at some distance in front of the clock.

Observations of the transit of "clock stars" across the meridian are made from time to time, and from these the rate of the clock is deduced in terms of "mean solar time."

The experimental pendulum is then set a swinging, and the two pendulums are observed through the telescope. Let us suppose that the time of a single vibration is not exactly that of the clock pendulum, but a little more.

The observer at the telescope sees the clock pendulum always gaining on the experimental pendulum, till at last the experimental pendulum just hides the white spot on the clock pendulum as it crosses the vertical line. The time at which this takes place is observed and recorded as the First Positive Coincidence.

The clock pendulum continues to gain on the other,

and after a certain time the two pendulums cross the vertical line at the same instant in opposite directions. The time of this is recorded as the First Negative Coincidence. After an equal interval of time there will be a second positive coincidence, and so on.

By this method the clock itself counts the number, N, of vibrations of its own pendulum between the coincidences. During this time the experimental pendulum has executed one vibration less than the clock. Hence the time of vibration of the experimental pendulum is $\dfrac{N}{N-1}$ seconds of clock time.

When there is no exact coincidence, but when the clock pendulum is ahead of the experimental pendulum at one passage of the vertical and behind at the next, a little practice on the part of the observer will enable him to estimate at what time between the passages the two pendulums must have been in the same phase. The epoch of coincidence can thus be estimated to a fraction of a second.

126. Estimation of Error

The experimental pendulum will go on swinging for some hours, so that the whole time to be measured may be ten thousand or more vibrations.

But the error introduced into the calculated time of vibration, by a mistake even of a whole second in noting the time of vibration, may be made exceedingly small by prolonging the experiment.

For if we observe the first and the nth coincidence, and find that they are separated by an interval of N seconds of the clock, the experimental pendulum will have lost n vibrations, as compared with the clock, and will have made $N-n$ vibrations in N seconds. Hence the time of a single vibration is $T=\dfrac{N}{N-n}$ seconds of clock time.

Let us suppose, however, that by a mistake of a

second we note down the last coincidence as taking place $N + 1$ seconds after the first. The value of T as deduced from this result would be

$$T' = \frac{N+1}{N+1-n}$$

and the error introduced by the mistake of a second will be

$$T' - T = \frac{N+1}{N+1-n} - \frac{N}{N-n}$$

$$= \frac{n}{(N+1-n)(N-n)}.$$

If N is 10,000 and n is 100, a mistake of one second in noting the time of coincidence will alter the value of T only about one-millionth part of its value.

CHAPTER VIII

UNIVERSAL GRAVITATION

127. Newton's Method

The most instructive example of the method of dynamical reasoning is that by which Newton determined the law of the force with which the heavenly bodies act on each other.

The process of dynamical reasoning consists in deducing from the successive configurations of the heavenly bodies, as observed by astronomers, their velocities and their accelerations, and in this way determining the direction and the relative magnitude of the force which acts on them.

Kepler had already prepared the way for Newton's investigation by deducing from a careful study of the observations of Tycho Brahe the three laws of planetary motion which bear his name.

128. Kepler's Laws

Kepler's Laws are purely kinematical. They completely describe the motions of the planets, but they say nothing about the forces by which these motions are determined.

Their dynamical interpretation was discovered by Newton.

The first and second laws relate to the motion of a single planet.

Law I.—The areas swept out by the vector drawn from the sun to a planet are proportional to the times of describing them.—If h denotes twice the area swept out in unit of time, twice the area swept out in time t will be ht, and if P is the mass of the planet, Pht will be the mass-area, as defined in Article 68. Hence the angular momentum of the planet about the sun, which is the rate of change of the mass-area, will be Ph, a constant quantity.

Hence, by Article 70, the force, if any, which acts on the planet must have no moment with respect to the sun, for if it had it would increase or diminish the angular momentum at a rate measured by the value of this moment.

Hence, whatever be the force which acts on the planet, the direction of this force must always pass through the sun.

129. Angular Velocity

Definition. The angular velocity of a vector is the rate at which the angle increases which it makes with a fixed vector in the plane of its motion.

If ω is the angular velocity of a vector, and r its length, the rate at which it sweeps out an area is $\frac{1}{2}\omega r^2$. Hence,

$$h = \omega r^2$$

and since h is constant, ω, the angular velocity of a planet's motion round the sun, varies inversely as the square of the distance from the sun.

This is true whatever the law of force may be, provided the force acting on the planet always passes through the sun.

130. Motion about the Centre of Mass

Since the stress between the planet and the sun acts on both bodies, neither of them can remain at rest. The only point whose motion is not affected by the stress is the centre of mass of the two bodies.

Fig. 15.

If r is the distance SP (fig. 15), and if C is the centre of mass, $SC = \dfrac{Pr}{S+P}$ and $CP = \dfrac{Sr}{S+P}$. The angular momentum of P about C is $P\omega \dfrac{S^2r^2}{(S+P)^2} = \dfrac{PS^2h}{(S+P)^2}$.

131. The Orbit

We have already made use of diagrams of configuration and of velocity in studying the motion of a material system. These diagrams, however, represent only the state of the system at a given instant; and this state is indicated by the relative position of points corresponding to the bodies forming the system.

It is often, however, convenient to represent in a single diagram the whole series of configurations or velocities which the system assumes. If we suppose the points of the diagram to move so as continually to represent the state of the moving system, each point of the diagram will trace out a line, straight or curved.

On the diagram of configuration, this line is called, in general, the Path of the body. In the case of the heavenly bodies it is often called the Orbit.

132. The Hodograph

On the diagram of velocity the line traced out by each moving point is called the Hodograph of the body to which it corresponds.

The study of the Hodograph, as a method of investigating the motion of a body, was introduced by Sir W. R. Hamilton. The hodograph may be defined as the path traced out by the extremity of a vector which continually represents, in direction and magnitude, the velocity of a moving body.

In applying the method of the hodograph to a planet, the orbit of which is in one plane, we shall find it convenient to suppose the hodograph turned round its origin through a right angle, so that the vector of the hodograph is perpendicular instead of parallel to the velocity it represents.

133. Kepler's Second Law

Law II.—The orbit of a planet with respect to the sun is an ellipse, the sun being in one of the foci.

Let $APQB$ (fig. 16) be the elliptic orbit. Let S be the sun in one focus, and let H be the other focus.

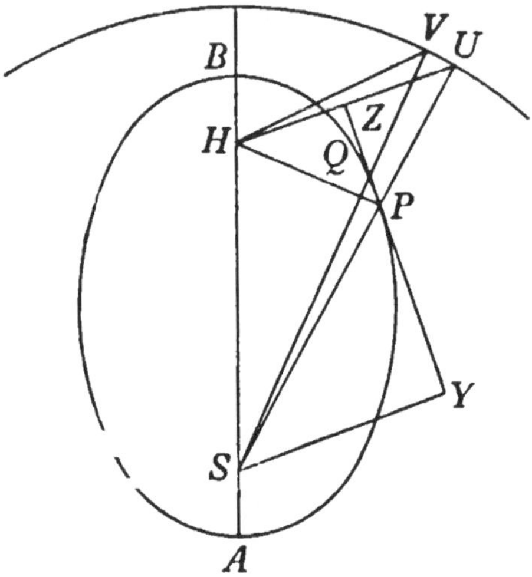

Fig. 16.

Produce SP to U, so that SU is equal to the transverse axis AB, and join HU, then HU will be proportional and perpendicular to the velocity at P.

For bisect HU in Z and join ZP; ZP will be a tangent to the ellipse at P; let SY be a perpendicular from S on this tangent.

If v is the velocity at P, and h twice the area swept out in unit of time, $h = vSY$.

Also if b is half the conjugate axis of the ellipse

$$SY.HZ = b^2.$$

Now $$HU = 2HZ;$$

hence $$v = \frac{1}{2}\frac{h}{b^2} HU.$$

Hence HU is always proportional to the velocity, and it is perpendicular to its direction. Now SU is always equal to AB. Hence the circle whose centre is S and radius AB is the hodograph of the planet, H being the origin of the hodograph.

The corresponding points of the orbit and the hodograph are those which lie in the same straight line through S.

Thus P corresponds to U and Q to V.

The velocity communicated to the body during its passage from P to Q is represented by the geometrical difference between the vectors HU and HV, that is, by the line UV, and it is perpendicular to this arc of the circle, and is therefore, as we have already proved, directed towards S.

If PQ is the arc described in [a very small] time, then UV represents the acceleration [of velocity in that time;] and since UV is on a circle whose centre is S, UV will be a measure of the angular [movement in that time] of the planet about S. Hence the acceleration is proportional to the angular velocity, and this by Art. 129 is inversely as the square of the distance SP. Hence the acceleration of the planet is in the direction of the sun, and is inversely as the square of the distance from the sun.

This, therefore, is the law according to which the attraction of the sun on a planet varies as the planet moves in its orbit and alters its distance from the sun.

134. Force on a Planet

As we have already shown, the orbit of the planet with respect to the centre of mass of the sun and planet has its dimensions in the ratio of S to $S+P$ to those of the orbit of the planet with respect to the sun.

If $2a$ and $2b$ are the axes of the orbit of the planet with respect to the sun, the area is πab, and if T is the time of going completely round the orbit, the value of h is $2\pi \frac{ab}{T}$. The velocity with respect to the sun is therefore $\pi \frac{a}{Tb} HU$.

With respect to the centre of mass it is

$$\frac{S}{S+P}\frac{\pi a}{Tb}HU.$$

The total acceleration of the planet towards the centre of mass [in describing an arc PQ] is

$$\frac{S}{S+P}\frac{\pi a}{Tb}UV$$

and the impulse on the planet whose mass is P is therefore

$$\frac{SP}{S+P}\frac{\pi a}{Tb}UV.$$

Let t be the time of describing PQ, then twice the area SPQ is

$$ht = \omega r^2 t$$

and $$UV = 2a\omega t = 2a\frac{h}{r^2}t = 4\pi\frac{a^2 b}{Tr^2}t.$$

Hence the force on the planet [being impulse divided by time] is

$$F = 4\pi^2\frac{SP}{S+P}\frac{a^3}{T^2 r^2}.$$

This then is the value of the stress or attraction between a planet and the sun in terms of their masses P and S, their mean distance a, their actual distance r, and the periodic time T.

135. Interpretation of Kepler's Third Law

To compare the attraction between the sun and different planets, Newton made use of Kepler's third law.

Law III.—The squares of the periodic times of different planets are proportional to the cubes of their mean distances. In other words $\frac{a^3}{T^2}$ is a constant, say $\frac{c}{4\pi^2}$.

Hence $$F = c\frac{SP}{S+P}\frac{1}{r^2}.$$

In the case of the smaller planets their masses are so small, compared with that of the sun, that $\frac{S}{S+P}$ may be put equal to 1, so that $F = c\frac{P}{r^2}$ or the attraction on a planet is proportional to its mass and inversely as the square of its distance.

136. Law of Gravitation

This is the most remarkable fact about the attraction of gravitation, that at the same distance it acts equally on equal masses of substances of all kinds. This is proved by pendulum experiments for the different kinds of matter at the surface of the earth. Newton extended the law to the matter of which the different planets are composed.

It had been suggested, before Newton proved it, that the sun as a whole attracts a planet as a whole, and the law of the inverse square had also been previously stated, but in the hands of Newton the doctrine of gravitation assumed its final form.

Every portion of matter attracts every other portion of matter, and the stress between them is proportional to the product of their masses divided by the square of their distance.

For if the attraction between a gramme of matter in the sun and a gramme of matter in a planet at distance r is $\frac{C}{r^2}$ where C is a constant, then if there are S grammes in the sun and P in the planet the whole attraction between the sun and one gramme in the planet will be $\frac{CS}{r^2}$, and the whole attraction between the sun and the planet will be $C\frac{SP}{r^2}$.

Comparing this statement of Newton's "Law of

Universal Gravitation" with the value of F formerly obtained we find

$$C\frac{SP}{r^2} = 4\pi^2\frac{SP}{S+P}\frac{a^3}{T^2r^2}$$

or

$$4\pi^2a^3 = C(S+P)T^2.$$

137. Amended Form of Kepler's Third Law

Hence Kepler's Third Law must be amended thus: The cubes of the mean distances are as the squares of the times multiplied into the sum of the masses of the sun and the planet.

In the case of the larger planets, Jupiter, Saturn, etc., the value of $S+P$ is considerably greater than in the case of the earth and the smaller planets. Hence the periodic times of the larger planets should be somewhat less than they would be according to Kepler's law, and this is found to be the case.

In the following table the mean distances (a) of the planets are given in terms of the mean distance of the earth, and the periodic times (T) in terms of the sidereal year:

Planet	a	T	a^3	T^2	$a^3 - T^2$
Mercury	0·387098	0·24084	0·0580046	0·0580049	−0·0000003
Venus	0·72333	0·61518	0·378451	0·378453	−0·000002
Earth	1·0000	1·00000	1·00000	1·00000	
Mars	1·52369	1·88082	3·53746	3·53747	−0·00001
Jupiter	5·20278	11·8618	140·832	140·701	+0·131
Saturn	9·53879	29·4560	867·914	867·658	+0·256
Uranus	19·1824	84·0123	7058·44	7058·07	+0·37
Neptune	30·037	164·616	27100·0	27098·4	+1·6

It appears from the table that Kepler's third law is very nearly accurate, for a^3 is very nearly equal to T^2, but that for those planets whose mass is less than that of the earth—namely, Mercury, Venus, and Mars—a^3 is less than T^2, whereas for Jupiter, Saturn, Uranus, and Neptune, whose mass is greater than that of the earth, a^3 is greater than T^2.

138. Potential Energy due to Gravitation

The potential energy of the gravitation between the bodies S and P may be calculated when we know the attraction between them in terms of their distance. The process of calculation by which we sum up the effects of a continually varying quantity belongs to the Integral Calculus, and though in this case the calculation may be explained by elementary methods, we shall rather deduce the potential energy directly from Kepler's first and second laws.

These laws completely define the motion of the sun and planet, and therefore we may find the kinetic energy of the system corresponding to any part of the elliptic orbit. Now, since the sun and planet form a conservative system, the sum of the kinetic and potential energies is constant, and therefore when we know the kinetic energy we may deduce that part of the potential energy which depends on the distance between the bodies.

139. Kinetic Energy of the System

To determine the kinetic energy we observe that the velocity of the planet with respect to the sun is by Article 133

$$v = \frac{1}{2}\frac{h}{b^2}HU.$$

The velocities of the planet and the sun with respect to the centre of mass of the system are respectively

$$\frac{S}{S+P}v \quad \text{and} \quad \frac{P}{S+P}v.$$

The kinetic energies of the planet and the sun are therefore

$$\tfrac{1}{2}P\frac{S^2}{(S+P)^2}v^2 \quad \text{and} \quad \tfrac{1}{2}S\frac{P^2}{(S+P)^2}v^2$$

and the whole kinetic energy is

$$\frac{1}{2}\frac{SP}{S+P}v^2 = \frac{1}{8}\frac{SP}{S+P}\frac{h^2}{b^4}HU^2.$$

To determine [more directly] v^2 in terms of SP or r, we observe that by the law of areas

$$vSY = h = \frac{2\pi ab}{T} \qquad \text{....(1)},$$

also by a property of the ellipse

$$HZ.SY = b^2 \qquad \text{. ....(2)},$$

and by the similar triangles HZP and SYP

$$\frac{SY}{HZ} = \frac{HP}{SP} = \frac{\mathrm{r}}{2a - \mathrm{r}} \qquad \text{....(3)};$$

multiplying (2) and (3) we find

$$SY^2 = \frac{b^2r}{2a - \mathrm{r}}.$$

Hence by (1)

$$v^2 = \frac{4\pi^2a^2b^2}{T^2}\,\frac{1}{SY^2} = \frac{4\pi^2a^2}{T^2}\left(\frac{2a}{r} - 1\right)$$

and the kinetic energy of the system is

$$\frac{4\pi^2a^3}{T^2}\,\frac{SP}{S+P}\left(\frac{1}{\mathrm{r}} - \frac{1}{2a}\right)$$

and this by the equation at the end of Article 136 becomes

$$C\,.\,SP\left(\frac{1}{r} - \frac{1}{2a}\right)$$

where C is the constant of gravitation.

This is the value of the kinetic energy of the two bodies S and P when moving [relatively] in an ellipse of which the transverse axis is $2a$.

140. Potential Energy of the System

The sum of the kinetic and potential energies is constant, but its absolute value is by Article 110 unknown, and not necessary to be known.

Hence if we [conclude, in accordance with the constancy of the total energy,] that the potential energy is of the form

$$K - C\,.\,SP\,\frac{1}{-}$$

the second term, which is the only one depending on the distance, r, is also the only one which we have anything to do with. The other term K represents the work done by gravitation while the two bodies originally at an infinite distance from each other are allowed to approach as near as their dimensions will allow them.

141. The Moon is a Heavy Body

Having thus determined the law of the force between each planet and the sun, Newton proceeded to show that the observed weight of bodies at the earth's surface and the force which retains the moon in her orbit round the earth are related to each other according to the same law of the inverse square of the distance.

This force of gravity acts in every region accessible to us, at the top of the highest mountains and at the highest point reached by balloons. Its intensity, as measured by pendulum experiments, decreases as we ascend; and although the height to which we can ascend is so small compared with the earth's radius that we cannot from observations of this kind infer that gravity varies inversely as the square of the distance from the centre of the earth, the observed decrease of the intensity of gravity is consistent with this law, the form of which had been suggested to Newton by the motion of the planets.

Assuming, then, that the intensity of gravity varies inversely as the square of the distance from the centre of the earth, and knowing its value at the surface of the earth, Newton calculated its value at the mean distance of the moon.

His first calculations were vitiated by his adopting an erroneous estimate of the dimensions of the earth. When, however, he had obtained a more correct value of this quantity* he found that the intensity of gravity

* And had demonstrated with great mathematical power the proposition assumed above, that the gravitation to a globe like the earth is exactly the same at all external points as if its mass were condensed to a point at its centre.

calculated for a distance equal to that of the moon was equal to the force required to keep the moon in her orbit.

He thus identified the force which acts between the earth and the moon with that which causes bodies near the earth's surface to fall towards the earth.

142. Cavendish's Experiment

Having thus shown that the force with which the heavenly bodies attract each other is of the same kind as that with which bodies that we can handle are attracted to the earth, it remained to be shown that bodies such as we can handle attract one another.

The difficulty of doing this arises from the fact that the mass of bodies which we can handle is so small compared with that of the earth, that even when we bring the two bodies as near as we can the attraction between them is an exceedingly small fraction of the weight of either.

We cannot get rid of the attraction of the earth, but we must arrange the experiment in such a way that it interferes as little as is possible with the effects of the attraction of the other body.

The apparatus devised by the Rev. John Michell* for this purpose was that which has since received the name of the Torsion Balance. Michell died before he was able to make the experiment, but his apparatus afterwards came into the hands of Henry Cavendish†, who improved it in many respects, and measured the attraction between [fixed] leaden balls and small balls suspended from the arms of the balance. A similar instrument was afterwards independently invented by Coulomb for measuring small electric and magnetic forces, and it continues to be the best instrument known to science for the measurement of small forces of all kinds.

* Of Queens' College, Cambridge, Woodwardian Professor of Geology, 1762–4. See *Memoir* by Sir A. Geikie, Cambridge, 1918.

† Of Peterhouse, Cambridge. See his *Scientific Writings*, 2 vols., Cambridge, 1920.

143. THE TORSION BALANCE*

The Torsion Balance consists of a horizontal rod suspended by a wire from a fixed support. When the rod is turned round by an external force in a horizontal plane it twists the wire, and the wire being elastic tends to resist this strain and to untwist itself. This force of torsion is proportional to the angle through which the wire is twisted, so that if we cause a force to act in a horizontal direction at right angles to the rod at its extremity, we may, by observing the angle through which the force is able to turn the rod, determine the magnitude of the force.

The force is proportional to the angle of torsion and to the fourth power of the diameter of the wire and inversely to the length of the rod and the length of the wire.

Hence, by using a long fine wire and a long rod, we may measure very small forces.

In the experiment of Cavendish two spheres of equal mass, m, are suspended from the extremities of the rod of the torsion balance. We shall for the present neglect the mass of the rod in comparison with that of the spheres. Two larger spheres of equal mass, M, are so arranged that they can be placed either at M and M or at M' and M'. In the former position they tend by their attraction on the smaller spheres, m and m, to turn the rod of the balance in the direction towards them. In the latter position they thus tend to turn it in the opposite direction. The torsion balance and its suspended spheres are enclosed in a case, to prevent their being disturbed by currents of air. The position of the rod of the balance is ascertained

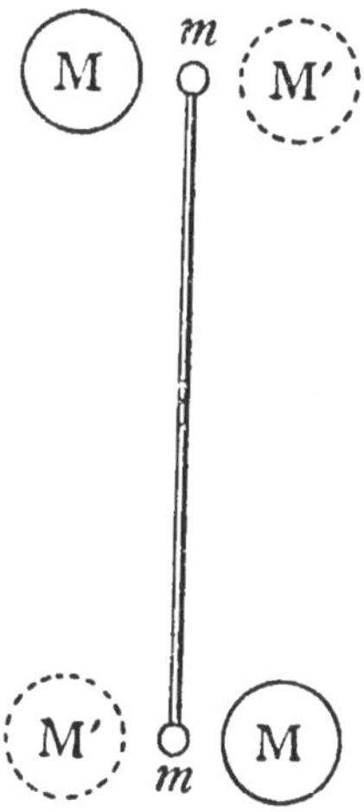

Fig. 17.

* See *infra*, p. 143.

by observing a graduated scale as seen by reflexion in a vertical mirror fastened to the middle of the rod. The balance is placed in a room by itself, and the observer does not enter the room, but observes the image of the graduated scale with a telescope.

144. Method of the Experiment

The time, T, of a double vibration of the torsion balance is first ascertained, and also the position of equilibrium of the centres of the suspended spheres.

The large spheres are then brought up to the positions M, M, so that the centre of each is at a distance from the position of equilibrium of the centre of the suspended sphere.

No attempt is made to wait till the vibrations of the beam have subsided, but the scale-divisions corresponding to the extremities of a single vibration are observed, and are found to be distant x and y respec-

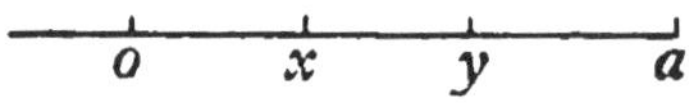

Fig. 18.

tively from the position of equilibrium. At these points the rod is, for an instant, at rest, so that its energy is entirely potential, and since the total energy is constant, the potential energy corresponding to the position x must be equal to that corresponding to the position y.

Now if T be the time of a double vibration about the point of equilibrium o, the potential energy due to torsion when the scale reading is x is by Article 119

$$\frac{2\pi^2 m}{T^2} x^2$$

and that due to the gravitation between m and M is by Article 140

$$K - C\frac{mM}{a - x}.$$

The potential energy of the whole system in the position x is therefore

$$K - C\frac{mM}{a-x} + \frac{2\pi^2 m}{T^2}x^2.$$

In the position y it is

$$K - C\frac{mM}{a-y} + \frac{2\pi^2 m}{T^2}y^2$$

and since the potential energy in these two positions is equal,

$$CmM\left(\frac{1}{a-y} - \frac{1}{a-x}\right) = \frac{2\pi^2 m}{T^2}(y^2 - x^2).$$

Hence
$$C = \frac{2\pi^2}{MT^2}(x+y)(a-x)(a-y).$$

By this equation C, the constant of gravitation, is determined in terms of the observed quantities, M the mass of the large spheres in grammes, T the time of a double vibration in seconds, and the distances x, y and a in centimetres.

According to Baily's experiments, $C = 6\cdot 5 \times 10^{-8}$. If we assume the unit of mass, so that at a distance unity it would produce an acceleration unity, the centimetre and the second being units, the unit of mass would be about $1\cdot 537 \times 10^7$ grammes, or 15·37 tonnes. This unit of mass reduces C, the constant of gravitation, to unity. It is therefore used in the calculations of physical astronomy.

145. Universal Gravitation

We have thus traced the attraction of gravitation through a great variety of natural phenomena, and have found that the law established for the variation of the force at different distances between a planet and the sun also holds when we compare the attraction between different planets and the sun, and also when we compare the attraction between the moon and the earth with that between the earth and heavy bodies at its surface. We have also found that the gravitation of equal masses at equal distances is the same whatever be the nature of the material of which the masses consist. This we ascertain by experiments on pendulums of different substances, and also by a comparison of the attraction of the sun on different planets, which are probably not alike in composition. The experiments of Baily* on spheres of different substances placed in the torsion balance confirm this law.

Since, therefore, we find in so great a number of cases occurring in regions remote from each other that the force of gravitation depends on the mass of bodies only, and not on their chemical nature or physical state, we are led to conclude that this is true for all substances.

For instance, no man of science doubts that two portions of atmospheric air attract one another, although we have very little hope that experimental methods will ever be invented so delicate as to measure or even to make manifest this attraction. But we know that there is attraction between any portion of air and the earth, and we find by Cavendish's experiment that gravitating bodies, if of sufficient mass, gravitate sensibly towards each other, and we conclude that two portions of air gravitate towards each other. But it is still extremely doubtful whether the medium of light and electricity is a gravitating substance, though it is certainly material and has mass†.

146. Cause of Gravitation

Newton, in his *Principia*, deduces from the observed motions of the heavenly bodies the fact that they attract one another according to a definite law.

* And more recently with extreme refinement by v. Jolly, Boys, Eötvös, and many others. Apparent weight is gravitation diminished by centrifugal reaction to the earth's rotation: if these did not vary in the same way for all kinds of matter, delicate weighings would detect the discrepancy: the experiments of Eötvös show that it could not exceed five parts in 10^8. See *infra*, p. 143.

† See *infra*, p. 144.

This he gives as a result of strict dynamical reasoning, and by it he shows how not only the more conspicuous phenomena, but all the apparent irregularities of the motions of these bodies are the calculable results of this single principle. In his *Principia* he confines himself to the demonstration and development of this great step in the science of the mutual action of bodies. He says nothing about the means by which bodies are made to gravitate towards each other. We know that his mind did not rest at this point—that he felt that gravitation itself must be capable of being explained, and that he even suggested an explanation depending on the action of an etherial medium pervading space. But with that wise moderation which is characteristic of all his investigations, he distinguished such speculations from what he had established by observation and demonstration, and excluded from his *Principia* all mention of the cause of gravitation, reserving his thoughts on this subject for the "Queries" printed at the end of his *Opticks*.

The attempts which have been made since the time of Newton to solve this difficult question are few in number, and have not led to any well-established result*.

147. Application of Newton's Method of Investigation

The method of investigating the forces which act between bodies which was thus pointed out and exemplified by Newton in the case of the heavenly bodies, was followed out successfully in the case of electrified and magnetized bodies by Cavendish, Coulomb, and Poisson.

The investigation of the mode in which the minute particles of bodies act on each other is rendered more difficult from the fact that both the bodies we consider

* See Appendix I, *infra*, p. 140.

and their distances are so small that we cannot perceive or measure them, and we are therefore unable to observe their motions as we do those of planets, or of electrified and magnetized bodies.

148. Methods of Molecular Investigations

Hence the investigations of molecular science have proceeded for the most part by the method of hypothesis, and comparison of the results of the hypothesis with the observed facts.

The success of this method depends on the generality of the hypothesis we begin with. If our hypothesis is the extremely general one that the phenomena to be investigated depend on the configuration and motion of a material system, then if we are able to deduce any available results from such an hypothesis, we may safely apply them to the phenomena before us*.

If, on the other hand, we frame the hypothesis that the configuration, motion, or action of the material system is of a certain definite kind, and if the results of this hypothesis agree with the phenomena, then, unless we can prove that no other hypothesis would account for the phenomena, we must still admit the possibility of our hypothesis being a wrong one.

149. Importance of General and Elementary Properties

It is therefore of the greatest importance in all physical inquiries that we should be thoroughly acquainted with the most general properties of material systems, and it is for this reason that in this book I have rather dwelt on these general properties than entered on the more varied and interesting field of the special properties of particular forms of matter.

* This is the subject of the next chapter.

[CHAPTER IX]

ON THE EQUATIONS OF MOTION OF A CONNECTED SYSTEM*

1. IN the fourth section of the second part of his *Mécanique Analytique*, Lagrange has given a method of reducing the ordinary dynamical equations of the motion of the parts of a connected system to a number equal to that of the degrees of freedom of the system.

The equations of motion of a connected system have been given in a different form by Hamilton, and have led to a great extension of the higher part of pure dynamics[1].

As we shall find it necessary, in our endeavours to bring electrical phenomena within the province of dynamics, to have our dynamical ideas in a state fit for direct application to physical questions, we shall devote this chapter to an exposition of these dynamical ideas from a physical point of view.

2. The aim of Lagrange was to bring dynamics under the power of the calculus. He began by expressing the elementary dynamical relations in terms of the corresponding relations of pure algebraical quantities, and from the equations thus obtained he deduced his final equations by a purely algebraical process. Certain quantities (expressing the reactions between the parts of the system called into play by its physical connexions) appear in the equations of motion of the component parts of the system, and Lagrange's investigation, as seen from a mathematical point of view, is a method of eliminating these quantities from the final equations.

In following the steps of this elimination the mind is exercised in calculation, and should therefore be kept

* This chapter, now added, is a reprint of Part IV, Chapter v. of Maxwell's *Treatise on Electricity and Magnetism* (1873).

[1] See Professor Cayley's "Report on Theoretical Dynamics," *British Association*, 1857; and Thomson and Tait's *Natural Philosophy* [1867].

free from the intrusion of dynamical ideas. Our aim, on the other hand, is to cultivate our dynamical ideas. We therefore avail ourselves of the labours of the mathematicians, and retranslate their results from the language of the calculus into the language of dynamics, so that our words may call up the mental image, not of some algebraical process, but of some property of moving bodies.

The language of dynamics has been considerably extended by those who have expounded in popular terms the doctrine of the Conservation of Energy, and it will be seen that much of the following statement is suggested by the investigation in Thomson and Tait's *Natural Philosophy*, especially the method of beginning with the theory of impulsive forces.

I have applied this method so as to avoid the explicit consideration of the motion of any part of the system except the coordinates or variables, on which the motion of the whole depends. It is doubtless important that the student should be able to trace the connexion of the motion of each part of the system with that of the variables, but it is by no means necessary to do this in the process of obtaining the final equations, which are independent of the particular form of these connexions.

The Variables

3. The number of degrees of freedom of a system is the number of data which must be given in order completely to determine its position. Different forms may be given to these data, but their number depends on the nature of the system itself, and cannot be altered.

To fix our ideas we may conceive the system connected by means of suitable mechanism with a number of moveable pieces, each capable of motion along a straight line, and of no other kind of motion. The imaginary mechanism which connects each of these pieces with the system must be conceived to be free from friction, destitute of inertia, and incapable of being strained by the action of the applied forces. The use of this mechanism is merely to assist the imagination in ascribing position, velocity, and momentum to what appear, in Lagrange's investigation, as pure algebraical quantities.

Let q denote the position of one of the moveable pieces as defined by its distance from a fixed point in its line of motion. We shall distinguish the values of q corresponding to the different pieces by the suffixes $_1$, $_2$, etc. When we are dealing with a set of quantities belonging to one piece only we may omit the suffix.

When the values of all the variables (q) are given, the position of each of the moveable pieces is known, and, in virtue of the imaginary mechanism, the configuration of the entire system is determined.

The Velocities

4. During the motion of the system the configuration changes in some definite manner, and since the configuration at each instant is fully defined by the values of the variables (q), the velocity of every part of the system, as well as its configuration, will be completely defined if we know the values of the variables (q), together with their velocities $\left(\frac{dq}{dt}, \text{ or, according to Newton's notation, } \dot{q}\right)$.

The Forces

5. *By* a proper regulation of the motion of the variables, any motion of the system, consistent with the nature of the connexions, may be produced. In order to produce this motion by moving the variable pieces, forces must be applied to these pieces.

We shall denote the force which must be applied to any variable q_r by F_r. The system of forces (F) is mechanically equivalent (in virtue of the connexions of the system) to the system of forces, whatever it may be, which really produces the motion.

The Momenta

6. When a body moves in such a way that its configuration, with respect to the force which acts on it, remains always the same (as, for instance, in the case of a force acting on a single particle in the line of its motion), the moving force is measured by the rate of increase of the momentum. If F is the moving force, and p the momentum,

$$F = \frac{dp}{dt},$$

whence $$p = \int F dt.$$

The time-integral of a force is called the Impulse of the force; so that we may assert that the momentum is the impulse of the force which would bring the body from a state of rest into the given state of motion.

In the case of a connected system in motion, the configuration is continually changing at a rate depending on the velocities ($\dot{q}$), so that we can no longer assume that the momentum is the time-integral of the force which acts on it.

But the increment δq of any variable cannot be greater than $\dot{q}'\delta t$, where δt is the time during which the increment takes place, and $\dot{q}'$ is the greatest value of the velocity during that time. In the case of a system moving from rest under the action of forces always in the same direction, this is evidently the final velocity.

If the final velocity and configuration of the system are given, we may conceive the velocity to be communicated to the system in a very small time δt, the original configuration differing from the final configuration by quantities δq_1, δq_2, etc., which are less than $\dot{q}_1\delta t$, $\dot{q}_2\delta t$, etc. respectively.

The smaller we suppose the increment of time δt, the greater must be the impressed forces, but the time-integral, or impulse, of each force will remain finite. The limiting value of the impulse, when the time is diminished and ultimately vanishes, is defined as the *instantaneous* impulse; and the momentum p, corresponding to any variable q, is defined as the impulse corresponding to that variable, when the system is brought instantaneously from a state of rest into the given state of motion.

This conception, that the momenta are capable of being produced by instantaneous impulses on the system at rest, is introduced only as a method of defining the magnitude of the momenta; for the momenta of the system depend only on the instantaneous state of motion of the system, and not on the process by which that state was produced.

In a connected system the momentum corresponding to any variable is in general a linear function of the velocities of all the variables, instead of being, as in the dynamics of a particle, simply proportional to the velocity.

The impulses required to change the velocities of the system suddenly from $\dot{q}_1$, $\dot{q}_2$, etc. to $\dot{q}_1'$, $\dot{q}_2'$, etc. are evidently equal to $p_1' - p_1$, $p_2' - p_2$, etc. the changes of momentum of the several variables.

Work done by a Small Impulse

7. The work done by the force F_1 during the impulse is the space-integral of the force, or

$$W = \int F_1 dq_1$$

$$= \int F_1 \dot{q}_1 dt.$$

If $\dot{q}_1'$ is the greatest and $\dot{q}_1''$ the least value of the velocity $\dot{q}_1$ during the action of the force, W must be less than

$$\dot{q}_1' \int F dt \text{ or } \dot{q}_1' (p_1' - p_1),$$

and greater than

$$\dot{q}_1'' \int F dt \text{ or } \dot{q}_1'' (p_1' - p_1).$$

If we now suppose the impulse $\int F dt$ to be diminished without limit, the values of $\dot{q}_1'$ and $\dot{q}_1''$ will approach and ultimately coincide with that of $\dot{q}_1$, and we may write $p_1' - p = \delta p_1$; so that the work done is ultimately

$$\delta W_1 = \dot{q}_1 \delta p_1,$$

or, *the work done by a very small impulse is ultimately the product of the impulse and the velocity.*

Increment of the Kinetic Energy

8. When work is done in setting a conservative system in motion, energy is communicated to it, and the system becomes capable of doing an equal amount of work against resistances before it is reduced to rest.

The energy which a system possesses in virtue of its motion is called its Kinetic Energy, and is communicated to it in the form of the work done by the forces which set it in motion.

If T be the kinetic energy of the system, and if it becomes $T + \delta T$, on account of the action of an infinitesimal impulse whose components are δp_1, δp_2, etc. the increment δT must be the sum of the quantities of work done by the components of the impulse, or in symbols,

$$\delta T = \dot{q}_1 \delta p_1 + \dot{q}_2 \delta p_2 + \text{etc.}$$
$$= \Sigma (\dot{q} \delta p). \qquad \ldots\ldots(1).$$

The instantaneous state of the system is completely defined if the variables and the momenta are given. Hence the kinetic energy, which depends on the instantaneous state of the system, can be expressed in terms of the variables (q), and the momenta (p). This is the mode of expressing T introduced by Hamilton. When T is expressed in this way we shall distinguish it by the suffix $_p$, thus, T_p.

The complete variation of T_p is

$$\delta T_p = \Sigma \left(\frac{\partial T_p}{\partial p} \delta p \right) + \Sigma \left(\frac{\partial T_p}{\partial q} \delta q \right) \ldots\ldots(2).$$

The last term may be written

$$\Sigma\left(\frac{\partial T_p}{\partial q}\dot{q}\,\delta t\right),$$

which diminishes with δt, and ultimately vanishes [compared with the first term] when the impulse becomes instantaneous.

Hence, equating the coefficients of δp in equations (1) and (2), we obtain

$$\dot{q} = \frac{\partial T_p}{\partial p} \qquad \text{.....(3)},$$

or, *the velocity corresponding to the variable q is the differential coefficient of T_p with respect to the corresponding momentum p.*

We have arrived at this result by the consideration of impulsive forces. By this method we have avoided the consideration of the change of configuration during the action of the forces. But the instantaneous state of the system is in all respects the same, whether the system was brought from a state of rest to the given state of motion by the transient application of impulsive forces, or whether it arrived at that state in any manner, however gradual.

In other words, the variables, and the corresponding velocities and momenta, depend on the actual state of motion of the system at the given instant, and not on its previous history.

Hence, the equation (3) is equally valid, whether the state of motion of the system is supposed due to impulsive forces, or to forces acting in any manner whatever.

We may now therefore dismiss the consideration of impulsive forces, together with the limitations imposed on their time of action, and on the changes of configuration during their action.

Hamilton's Equations of Motion

9. We have already shown that

$$\frac{\partial T_p}{\partial p} = \dot{q} \qquad \text{.....(4)}.$$

Let the system move in any arbitrary way, subject to the conditions imposed by its connexions; then the variations of p and q are

$$\delta p = \frac{dp}{dt}\delta t, \qquad \delta q = \dot{q}\,\delta t \qquad \text{......(5)}.$$

Hence

$$\frac{\partial T_p}{\partial p}\delta p = \frac{dp}{dt}\dot{q}\,\delta t$$

$$= \frac{dp}{dt}\delta q \qquad \text{.....(6)},$$

and the complete variation of T_p is

$$\delta T_p = \Sigma\left(\frac{\partial T_p}{\partial p}\delta p + \frac{\partial T_p}{\partial q}\delta q\right)$$

$$= \Sigma\left\{\left(\frac{dp}{dt} + \frac{\partial T_p}{\partial q}\right)\delta q\right\} \qquad \text{......(7)}.$$

But the increment of the kinetic energy arises from the work done by the impressed forces, or

$$\delta T_p = \Sigma\,(F\,\delta q) \qquad \text{......(8)}.$$

In these two expressions the variations δq are all independent of each other, so that we are entitled to equate the coefficients of each of them in the two expressions (7) and (8). We thus obtain

$$F_r = \frac{dp_r}{dt} + \frac{\partial T_p}{\partial q_r} \qquad \text{......(9)},$$

where the momentum p_r and the force F_r belong to the variable q_r*.

There are as many equations of this form as there are variables. These equations were given by Hamilton. They show that the force corresponding to any variable is the sum of two parts. The first part is the rate of increase of the momentum of that variable with respect to the time. The second part is the rate of increase of the kinetic energy per unit of increment of the variable, the other variables and all the momenta being constant.

* But see *infra*, p. 158.

The Kinetic Energy expressed in Terms of the Momenta and Velocities

10. Let p_1, p_2, etc. be the momenta, and $\dot{q}_1$, $\dot{q}_2$, etc. the velocities at a given instant, and let p_1, p_2, etc., $\dot{\mathrm{q}}_1$, $\dot{\mathrm{q}}_2$, etc. be another system of momenta and velocities, such that

$$\mathrm{p}_1 = np_1, \quad \dot{\mathrm{q}}_1 = n\dot{q}_1, \text{ etc.} \qquad \ldots\ldots(10).$$

It is manifest that the systems p, $\dot{\mathrm{q}}$ will be consistent with each other if the systems p, $\dot{q}$ are so.

Now let n vary by δn. The work done by the force F_1 is [by § 7]

$$F_1\delta \mathrm{q}_1 = \dot{\mathrm{q}}_1\delta\mathrm{p}_1 = \dot{q}_1 p_1 n\,\delta n \qquad \ldots\ldots(11).$$

Let n increase from 0 to 1; then the system is brought from a state of rest into the state of motion $(\dot{q}p)$, and the whole work expended in producing this motion is

$$(\dot{q}_1 p_1 + \dot{q}_2 p_2 + \text{etc.}) \int_0^1 n\,dn \qquad \ldots\ldots(12).$$

But

$$\int_0^1 n\,dn = \tfrac{1}{2},$$

and the work spent in producing the motion is equivalent to the kinetic energy. Hence

$$T_{p\dot{q}} = \tfrac{1}{2}(p_1\dot{q}_1 + p_2\dot{q}_2 + \text{etc.}) \qquad \ldots\ldots(13),$$

where $T_{p\dot{q}}$ denotes the kinetic energy expressed in terms of the momenta and velocities. The variables q_1, q_2, etc., do not enter into this expression.

The kinetic energy is therefore half the sum of the products of the momenta into their corresponding velocities.

When the kinetic energy is expressed in this way we shall denote it by the symbol $T_{p\dot{q}}$. It is a function of the momenta and velocities only, and does not involve the variables themselves.

11. There is a third method of expressing the kinetic energy, which is generally, indeed, regarded as the fundamental one. By solving the equations (3) we may express the momenta in terms of the velocities, and

then, introducing these values in (13), we shall have an expression for T involving only the velocities and the variables. When T is expressed in this form we shall indicate it by the symbol $T_{\dot{q}}$. This is the form in which the kinetic energy is expressed in the equations of Lagrange.

12. It is manifest that, since T_p, $T_{\dot{q}}$, and $T_{p\dot{q}}$, are three different expressions for the same thing,

$$T_p + T_{\dot{q}} - 2T_{p\dot{q}} = 0,$$

or
$$T_p + T_{\dot{q}} - p_1\dot{q}_1 - p_2\dot{q}_2 - \text{etc.} = 0 \quad \ldots(14).$$

Hence, if all the quantities p, q, and $\dot{q}$ vary,

$$\left(\frac{\partial T_p}{\partial p_1} - \dot{q}_1\right)\delta p_1 + \left(\frac{\partial T_p}{\partial p_2} - \dot{q}_2\right)\delta p_2 + \text{etc.}$$
$$+\left(\frac{\partial T_{\dot{q}}}{\partial \dot{q}_1} - p_1\right)\delta \dot{q}_1 + \left(\frac{\partial T_{\dot{q}}}{\partial \dot{q}_2} - p_2\right)\delta \dot{q}_2 + \text{etc.}$$
$$+\left(\frac{\partial T_p}{\partial q_1} + \frac{\partial T_{\dot{q}}}{\partial q_1}\right)\delta q_1 + \left(\frac{\partial T_p}{\partial q_2} + \frac{\partial T_{\dot{q}}}{\partial q_2}\right)\delta q_2 + \text{etc.} = 0 \quad (15).$$

The variations δp are not independent of the variations δq and $\delta\dot{q}$, so that we cannot at once assert that the coefficient of each variation in this equation is zero. But we know, from equations (3), that

$$\frac{\partial T_p}{\partial p_1} - \dot{q}_1 = 0, \text{ etc.} \quad \ldots\ldots(16),$$

so that the terms involving the variations δp vanish of themselves.

The remaining variations $\delta\dot{q}$ and δq are now all independent*, so that we find, by equating to zero the coefficients of $\delta\dot{q}_1$, etc.,

$$p_1 = \frac{\partial T_{\dot{q}}}{\partial \dot{q}_1}, \quad p_2 = \frac{\partial T_{\dot{q}}}{\partial \dot{q}_2}, \text{ etc.} \quad \ldots\ldots(17);$$

or, *the components of momentum are the differential coefficients of* $T_{\dot{q}}$ *with respect to the corresponding velocities.*

* See *infra*, p. 159.

Again, by equating to zero the coefficients of δq_1, etc.,

$$\frac{\partial T_p}{\partial q_1} + \frac{\partial T_{\dot q}}{\partial q_1} = 0 \qquad \text{......(18);}$$

or, *the differential coefficient of the kinetic energy with respect to any variable q_1 is equal in magnitude but opposite in sign when T is expressed as a function of the velocities instead of as a function of the momenta.*

In virtue of equation (18) we may write the equation of motion (9)

$$F_1 = \frac{dp_1}{dt} - \frac{\partial T_{\dot q}}{\partial q_1} \qquad \text{......(19),}$$

or

$$F_1 = \frac{d}{dt}\frac{\partial T_{\dot q}}{\partial \dot q_1} - \frac{\partial T_{\dot q}}{\partial q_1} \qquad \text{......(20),}$$

which is the form in which the equations of motion were given by Lagrange.

13. In the preceding investigation we have avoided the consideration of the form of the function which expresses the kinetic energy in terms either of the velocities or of the momenta. The only explicit form which we have assigned to it is

$$T_{p\dot q} = \tfrac{1}{2}(p_1\dot q_1 + p_2\dot q_2 + \text{etc.}) \qquad \text{......(21),}$$

in which it is expressed as half the sum of the products of the momenta each into its corresponding velocity.

We may express the velocities in terms of the differential coefficients of T_p with respect to the momenta, as in equation (3) [; thus]

$$T_p = \frac{1}{2}\left(p_1\frac{\partial T_p}{\partial p_1} + p_2\frac{\partial T_p}{\partial p_2} + \text{etc.}\right) \quad \text{...(22).}$$

This shows that T_p is a homogeneous function of the second degree of the momenta p_1, p_2, etc.

We may also express the momenta in terms of $T_{\dot q}$, and we find

$$T_{\dot q} = \frac{1}{2}\left(\dot q_1\frac{\partial T_{\dot q}}{\partial \dot q_1} + \dot q_2\frac{\partial T_{\dot q}}{\partial \dot q_2} + \text{etc.}\right) \text{......(23)}$$

which shows that $T_{\dot{q}}$ is a homogeneous function of the second degree with respect to the velocities $\dot{q}_1$, $\dot{q}_2$, etc.

If we write

$$P_{11} \text{ for } \frac{\partial^2 T_{\dot{q}}}{\partial \dot{q}_1{}^2}, \qquad P_{12} \text{ for } \frac{\partial^2 T_{\dot{q}}}{\partial \dot{q}_1 \partial \dot{q}_2}, \text{ etc.}$$

and
$$Q_{11} \text{ for } \frac{\partial^2 T_p}{\partial p_1{}^2}, \qquad Q_{12} \text{ for } \frac{\partial^2 T_p}{\partial p_1 \partial p_2}, \text{ etc.};$$

then, since both $T_{\dot{q}}$ and T_p are functions of the second degree of $\dot{q}$ and of p respectively, both the P's and the Q's will be functions of the variables q only, and independent of the velocities and the momenta. We thus obtain the expressions for T,

$$2T_{\dot{q}} = P_{11}\dot{q}_1{}^2 + 2P_{12}\dot{q}_1\dot{q}_2 + \text{etc.} \quad \ldots(24),$$

$$2T_p = Q_{11}p_1{}^2 + 2Q_{12}p_1p_2 + \text{etc.} \quad \ldots(25).$$

The momenta are expressed in terms of the velocities by the linear equations

$$p_1 = P_{11}\dot{q}_1 + P_{12}\dot{q}_2 + \text{etc.} \quad \ldots\ldots(26),$$

and the velocities are expressed in terms of the momenta by the linear equations

$$\dot{q}_1 = Q_{11}p_1 + Q_{12}p_2 + \text{etc.} \quad \ldots\ldots(27).$$

In treatises on the dynamics of a rigid body, the coefficients corresponding to P_{11}, in which the suffixes are the same, are called Moments of Inertia, and those corresponding to P_{12}, in which the suffixes are different, are called Products of Inertia. We may extend these names to the more general problem which is now before us, in which these quantities are not, as in the case of a rigid body, absolute constants, but are functions of the variables q_1, q_2, etc.

In like manner we may call the coefficients of the form Q_{11} Moments of Mobility, and those of the form Q_{12}, Products of Mobility. It is not often, however, that we shall have occasion to speak of the coefficients of mobility.

14. The kinetic energy of the system is a quantity essentially positive or zero. Hence, whether it be expressed in terms of the velocities, or in terms of the momenta, the coefficients must be such that no real values of the variables can make T negative.

There are thus a set of necessary conditions which the values of the coefficients P must satisfy. These conditions are as follows:

The quantities P_{11}, P_{22}, etc. must all be positive.

The $n - 1$ determinants formed in succession from the determinant

$$\begin{vmatrix} P_{11}, & P_{12}, & P_{13}, \ldots\ldots & P_{1n} \\ P_{12}, & P_{22}, & P_{23}, \ldots\ldots & P_{2n} \\ P_{13}, & P_{23}, & P_{33}, \ldots\ldots & P_{3n} \\ \cdots & \cdots & \cdots & \cdots \\ P_{1n}, & P_{2n}, & P_{3n}, \ldots\ldots & P_{nn} \end{vmatrix}$$

by the omission of terms with suffix 1, then of terms with either 1 or 2 in their suffix, and so on, must all be positive.

The number of conditions for n variables is therefore $2n - 1$.

The coefficients Q are subject to conditions of the same kind.

15. In this outline of the fundamental principles of the dynamics of a connected system, we have kept out of view the mechanism by which the parts of the system are connected. We have not even written down a set of equations to indicate how the motion of any part of the system depends on the variation of the variables. We have confined our attention to the variables, their velocities and momenta, and the forces which act on the pieces representing the variables. Our only assumptions are, that the connexions of the system are such that the time is not explicitly contained in the equations of condition, and that the principle of the conservation of energy is applicable to the system.

Such a description of the methods of pure dynamics is not unnecessary, because Lagrange and most of his followers, to whom we are indebted for these methods, have in general confined themselves to a demonstration of them, and, in order to devote their attention to the symbols before them, they have endeavoured to banish all ideas except those of pure quantity, so as not only to dispense with diagrams, but even to get rid of the ideas of velocity, momentum, and energy, after they have been once for all supplanted by symbols in the original equations. In order to be able to refer to the results of this analysis in ordinary dynamical language, we have endeavoured to retranslate the principal equations of the method into language which may be intelligible without the use of symbols.

As the development of the ideas and methods of pure mathematics has rendered it possible, by forming a mathematical theory of dynamics, to bring to light many truths which could not have been discovered without mathematical training*, so, if we are to form dynamical theories of other sciences, we must have our minds imbued with these dynamical truths as well as with mathematical methods.

In forming the ideas and words relating to any science, which, like electricity, deals with forces and their effects, we must keep constantly in mind the ideas appropriate to the fundamental science of dynamics, so that we may, during the first development of the science, avoid inconsistency with what is already established, and also that when our views become clearer, the language we have adopted may be a help to us and not a hindrance.

* It has also generalized our conception of dynamics, so that it is possible to assert that a physical system is of dynamical type although we may not have been able to form an idea of the configurations and motions that are represented by the variables. See Appendix II.

APPENDIX I (1920)

The Relativity of the Forces of Nature

THE idea of the forces of nature was introduced into science in definite form by Sir Isaac Newton, in the expression of his Laws of Motion in the Introduction to the *Principia*. He specified physical force as recognized and measured by the rate at which the velocity of the body on which it acts is changing with the time. This was the simplest measure conceivable; it was postulated tacitly that the forces so recognized correspond to actual invariant causes of motion, which are always present, in accordance with the uniformity of nature, whenever the same conditions of the surrounding system of bodies recur. An underlying question is thus suggested as to why this particular measure corresponds to objective nature, and not some more complex one, involving for example the velocity also, or the rate of change of the acceleration as well as that of the velocity.

But this introduction of the idea of forces of nature also gave rise to the practical need of specifying some definite mode of prescribing velocity and its rate of change. Position and velocity belong to one system of bodies in space and time, but are *relative* to some other system. The simplest plan is to postulate some standard system for general reference. Accordingly Newton laid down a scheme of absolute space and absolute time, with respect to which the movements and forces in nature are to be determined. It is then necessary for dynamical science to determine this scheme of reference provisionally, for the set of problems in hand, and continually to correct its specification as the advance of knowledge requires. Thus for ordinary purposes the space referred to the surrounding landscape and the time of an ordinary vibrator will suffice for a standard;

but in wider problems when the rotation of the Earth has to be recognized these are no longer adequate, and must be replaced by a scheme of space and time which does not revolve with the Earth; and so on. The revolution effected by Copernicus, in transferring the centre of reference from the Earth to the Sun, was thus a preliminary to this dynamical order of ideas. We can conceive an ultimate system of space and time as that frame to which the stars and stellar universes can be related, so as to secure the greatest simplicity in the mode of describing their motions. Any frame of space and time to which the forces of nature are thus consistently referred, with sufficient precision for the purposes in view, has been named a frame of inertia, because with respect to it these forces are determined by the Newtonian product inertia-acceleration. For ordinary purposes there are many equally approximate frames of inertia; any uniform motion of translation of such a frame will make no difference in its practical efficacy.

This postulation of a standard space and a standard time in the *Principia* in 1687 was made with a view to simple treatment of the motions of the planetary bodies in space: but it at once excited the criticism of philosophers both at home and abroad, though apparently they had no practical alternative to offer. The illustrious Leibniz continued to challenge its validity; his epistolary controversy with Dr Samuel Clarke, who assumed on abstract principles the championship of the Newtonian practical formulation, is one of the classics of metaphysical philosophy. Our own Berkeley as a student at Trinity College, Dublin, where he was already thinking out his critical idealist scheme of philosophy, came up against the same kind of difficulties in his study of the foundations of the Newtonian system of the world. Have we any warrant for assigning an absolute frame of space and time for the laws of nature, especially with respect to the vast vacant spaces of astronomy? and could we have valid means of recognizing any such frame? It is perhaps largely a question of expression; if philosophers could come to mean the same thing by the terms they use they ought to agree, otherwise the universal validity of the operations of the mind might come into doubt.

The validity of such practical specification of a standard space and time has remained abstractly an open question; in recent years it has again come prominently into discussion. The phenomena of electricity and light had been thoroughly explained, under the guidance of Faraday and Clerk Maxwell, in terms of activities established and propagated in an aether of space, which is at rest in undisturbed regions so that it is natural to fit into it the Newtonian frame of space and time. The aether would thus be space and time endowed with physical properties, inertia and elasticity, as well as properties of extension. But it was found later that very refined and delicate experiments that seemed qualified to determine the motion of the Earth relative to the aether—and it must be at least of the order of its orbital velocity round the Sun—all failed to show any result. This was not unexpected, and was in fact quite explicable on the lines of Maxwell's theory. But it has stimulated independent trains of thought which in the end have propounded the question whether it is possible, at the cost of more complex and provisional modes of reference, to get rid altogether of the universal forces of nature such as gravitation, whose sole evidence is the acceleration of motions for which they are introduced as the cause. Thus if the scale of time is made to alter from place to place, so that duration is a function of position, the apparent values of gravitational accelerations will of course all be changed. The argument then is that (cf. § 103) all bodies in the same locality possess exactly the same acceleration on account of gravitation: if this universal feature can be absorbed into a complex reckoning of space and time, and so got rid of, the other relations of physical nature will merely have to become relative to the slightly altered reckoning introduced for this purpose. But our knowledge of physical extension and duration comes mainly from the sense of sight: little of it would have been acquired by a race without vision. It is impossible to ignore the rays of light as messengers of direction and duration from all parts of the visible universe. These essential and determining phenomena of radiation also must become mere local features of time and space, or else they would put us in connexion with a universal frame with respect to which they are propagated. However that may be, a theory which claims to be founded on metaphysical principles has recently been developed by Einstein and a numerous and important school, in which it is found that the forces of gravitation, and no other, can be represented with precision as inherent in a more complicated scheme of space and time instead of in the physical nature that that frame helps to describe; while at the same time they thereby fall into line with the electrodynamic doctrines of relativity above-mentioned.

It has been recognized however also that the same results can arise naturally, and without involving revolutionary ideas of time and space, as a slight (though analytically complex) expansion of the fundamental physical formulation of Least Action (*infra*); the special relations of stress, energy, and momentum on which as criteria the theory had to develop being in fact already implicit on that universal principle.

This alteration in the mode of expression of Newtonian gravitation of course makes very little practical difference; it however claimed special notice as removing one outstanding slight discrepancy with observation, in the motion of the inner planet Mercury, which had previously to be ascribed to an assumed distribution of mass between the planet and the Sun. Such an equivalent warping of the frame of space and time must also affect either in reality or in appearance the propagation of radiation wherever gravitation is intense. One such inference is that rays of light would be very slightly deviated in passing close to the Sun: and the results of the Greenwich and Cambridge astronomers who observed the solar eclipse of 1919 have in fact confirmed the required amount rather closely. But another result of such an order of ideas, of a spectroscopic character, still lacks any definite confirmation.

The primary *desideratum* as regards gravitation was to find a mathematical mode of expression which would bring it into touch with the theory of electrical agencies and of radiation, from which it had been isolated, and even, as regards the nature of the relation of inertia to weight, in very slight discord. This has been done by ascribing the acceleration common to all bodies merely to an altering frame of reference, instead of the introduction into ordinary space of an intrinsic gravitational potential function indicating an independent type of local activity. For velocities very large, thousands of times greater than the actual speeds of the heavenly bodies, the results of this view would be quite different from the simple Newtonian gravitation, and with our means of expression they would be of extreme complication: but in the actual stellar world the difference is excessively slight, and in the right direction. So far from replacing Newtonian astronomy it can only establish connexion with reality by making use of its representations and methods. We may perhaps conclude that the linking up of gravitation, previously isolated, with other physical agencies has been effected: but we ought not to exclude a hope that the mode of expression of this connexion may in time be greatly simplified, especially by more attention to the Principle of Action, as it is only very small changes that are involved. Meantime the extrapolation, based on the present general formulation of the theory, to exploration of universes involving far higher speeds than the stars possess in our own, is a fascinating subject for abstruse mathematical speculation.

The general doctrine of relativity, at any rate in its more extreme formulations, impugns the validity of arguments such as those of §§ 105–6. This question must relate to the meanings of the parties to a controversy. If we were shut off from sight of the stars there might be greater reason for claiming that it would be unphilosophic even to mention such a thing as an absolute rotation of the Earth, or any movement that could not be expressed as conditioned by adjacent bodies. That type of theory claims to settle all things by local scale and clock: but it also has in practice to requisition the use of the directions and periodic times of rays of light as valid means of discrimination. Unless the rays are to bend to the control of scale and clock, these measures will not be concordant: if they do, the connexion may be held to fix the frame with respect to which the rays travel with their assumed universal velocity, and thus to determine in part what has been regarded as the aether of space. An artificial gravitational field could be simulated by accelerating the frame of reference, provided it is not done by a mere algebraic change of coordinates: but the rays of light might have different speeds in it forward and backward, which would seem to involve a discriminating criterion for any such unrestricted "principle of equivalence" of a gravitational field to a changing frame. Any purely algebraic theory is an abstraction from the wider field of phenomena, and an essential question for it is the range of its own validity.

Note to § 145

As the mean result of numerous modern determinations Cavendish's value 5·45 for the mean density of the Earth has to be increased by less than two per cent. The torsion apparatus has been very greatly reduced in size and improved by C. V. Boys (1894) by use of his extremely fine and perfectly elastic quartz fibres for the torsional suspension.

The Michell-Cavendish-Coulomb torsion balance has been applied by Eötvös to test the proportionality of gravitation to mass, with results of extreme precision. The apparent weight of a body is its gravitation to the Earth as modified by a centrifugal force which is oblique to the vertical, being directed away from the axis of the diurnal rotation. The latter part is of course considerable, being a fraction of one per cent. of the whole; and it has a horizontal component along the meridian. If the mass factors in the two parts were not exactly equal a torsion balance, with the ends of its horizontal bar loaded by masses of different substances, would indicate a deflection of the bar relative to its frame when it is turned round the vertical from east-west to west-east. Eötvös (1891, 1897) thus found that any defect of proportionality of weight to mass must actually be less than one part in twenty million: and Zeeman, by a reduced apparatus with quartz-fibre suspension, has recently (1917) pushed the result still lower and extended it to crystals and to substances of radioactive origin. As it happens, this is nearly to the same order as the optical and electric verifications of absence of effects of convection through the aether owing to the Earth's motion.

If m is the inertia-mass of the centrifugal force and m' the mass which gravitates, then if m were equal to m' the apparent weight would be in the same direction for all substances and the experiment would show no result. Any possible result would thus be readily computed as that due to the centrifugal force of the excess $m-m'$, the moment of its horizontal component round the axis of torsion operating different ways in the two positions of the bar and frame.

It is a consequence of Maxwell's electrodynamics that when a body loses energy ϵ by radiation it loses inertia of amount ϵ/c^2, where c is the velocity of light. In modern extensions of that theory all energy has inertia. The inertia of an electron seems to be all associated with its steady kinetic energy of motion. The closeness of the Eötvös result thus carries the conclusion that the inertia of an electron must all gravitate, and in fact that all energy possesses inertia which is also gravitative. Thus neither inertia nor gravitation could continue to be specific constants of matter: they must be connected up either with the aether in which matter subsists, or with the abstract reference-frame of space-time which is all that can remain if such a medium is denied.

APPENDIX II (1920)

The Principle of Least Action

THE great *desideratum* for any science is its reduction to the smallest number of dominating principles. This has been effected for dynamical science mainly by Sir William Rowan Hamilton, of Dublin (1834–5), building on the analytical foundations provided by Lagrange in the formulation of Least Action in terms of the methods of his Calculus of Variations (1758), and later (1788) but less fundamentally for physical purposes on the principle of virtual work in the *Mécanique Analytique*.

The principle of the Conservation of Energy, inasmuch as it can provide only one equation, cannot determine by itself alone the orbit of a single body, much less the course of a more complex system (thus §§ 107–112 above need some qualification). But if the body starts on its path from a given position in the field of force and with assigned velocity, the principle of energy then determines the velocity this body must have when it arrives at any other position, either in the course of free motion or under guidance by constraints such as are frictionless and so consume no energy. If W, a function of position, represents the potential energy of a body in the field, per unit mass, the velocity v of the body is in fact determined by the equation

$$\tfrac{1}{2}mv^2 + mW = \tfrac{1}{2}mv_0{}^2 + mW_0 = mE,$$

where the subscripts in v_0 and W_0 refer to the initial position; and mE is the total energy of the body in relation to the field of force, which is conserved throughout its path. Thus

$$v = (2E - 2W)^{\frac{1}{2}};$$

so that the velocity v depends, through W, on position alone.

Now we can propound the following problem. By

what path must the body, of mass m, be guided under frictionless constraint from an initial position A to a final position B in space, with given conserved total energy mE, so that the Action in the path, defined as the limit of the sum $\Sigma mv\delta s$, that is as $\int mvds$, where δs is an element of length of this path, shall, over each stage, be least possible? The method of treating the simpler problems of this kind is known to have been familiar to Newton: in the case of the present question, first vaguely proposed by Maupertuis* when President of the Berlin Academy under Frederic the Great, the solution was gradually evolved and enlarged by the famous Swiss mathematical family of Bernoulli and their compatriot Euler: and finally, extended to more complex cases, it gave rise, after Euler's treatise of date 1744, in the hands of the youthful Lagrange (*Turin Memoirs*, 1758) to the Calculus of Variations, the most fruitful expansion of the processes of the infinitesimal calculus, for purposes of physical science, since the time of Newton and Leibniz.

Let us draw in the given field of force a series of closely consecutive surfaces of constant velocity, and therefore of constant potential energy mW: and let us consider an orbit $ABCD$... intersecting these surfaces at the points $B, C, D, \ldots$. We shall regard, in the Newtonian manner†, the velocity as constant, say v_1, in the infinitesimal path from B to C, and constant, say v_2, from C to D: these elements of the path are thus to be regarded as straight, the field of force being supposed to operate by a succession of very slight impulses at B, C, D, ... such as in the limit, as the elements of the path diminish indefinitely, will converge to the continuous operation of a finite force.

* The notion of an Action possibly with minimal quality, not merely passive inertia, as concerned in the transmission of Potentia or energy, is ascribed to Leibniz by Helmholtz in 1887.

† Cf. *Principia*, Book I, Sec. II, Prop. I, on equable description of areas in a central orbit.

If $\Sigma v\delta s$ is to be a minimum over this section $ABCD$... of the path, then by the usual criterion any slight alteration, by frictionless constraint, which would compel the body to take locally an adjacent course $BC'D$, ought not to alter the value of the Action so far as regards the first order of small quantities. Now, on our representation of the force as a rapid succession of small impulses, the change so produced in the value of this function of Action is equal to

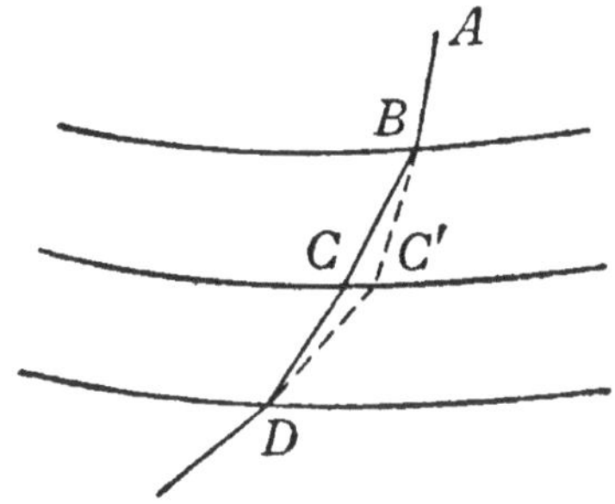

$$v_1 (BC' - BC) + v_2 (C'D - CD);$$

hence this must vanish, up to the first order. But $BC' - BC$ is equal to $-$ CC′ cos BCC', and $C'D - CD$ is equal to $-$ CC′ cos DCC'. Thus the condition for a stationary value is that the component of v_1 along CC′ is equal to the component of v_2 along the same direction, where CC′ is any element of length on the surface of constant v, that is of constant W, drawn through C. This involves that the impulse which must be imparted to the body at C in order to change its velocity from v_1 to v_2 must be wholly transverse to this surface: or, on passing to the limit, that the force acting on the body must everywhere be in the direction of the gradient of the potential W. That is, whatever the form of this potential function may be, the succession of impulses must be in the direction of its force; it is already prescribed by the form of v that they are of the amounts necessary to make changes in the velocity that are in accord with conservation of energy. These are just the criteria for a free orbit. Hence for any short arc of any free orbit the Action $m\Sigma v\delta s$ is smaller than it could be if the orbit were slightly altered locally owing to any frictionless constraint. The free orbit is thus describable as the path of advance that would be determined by minimum expenditure of Action in each stage, as the body proceeds: though this does not imply that the total expenditure of Action from one end to the other of a longer path is necessarily or always the least possible. This formula of stationary (or say *minimal*) Action, expressed by the variational equation

$$\delta\int mvds = 0, \text{ where } \tfrac{1}{2}mv^2 + mW = mE,$$

is by itself competent to select the actual free orbit from among all possible constrained paths.

And generally, for any dynamical system having kinetic energy expressed by a function T of a sufficient number of geometrical coördinates, and potential energy expressed by W, it can be shown that the course of motion from one given configuration to another is completely determined by the single variational equation

$$\delta\int Tdt = 0 \text{ subject to } T + W = E,$$

E being the total energy, which is prescribed as conserved, so that the variations contemplated in the motion must be due only to frictionless constraints.

Another form of the principle is that

$$\delta\int (T - W)\,dt = 0$$

provided the total time of motion from the given initial to the given final configuration is kept constant. This form is more convenient for analytical purposes because the mode of variation is not restricted to frictionless constraint; as conservation of the energy is not imposed, extraneous forces, which can be included in a modification of W, may be in operation imparting energy to the system. Constancy of the time of transit, which here takes the place of conservation of the energy, is analytically, though not physically, a simpler form of restriction. From this form the complete set of general equations of motion developed by Lagrange (see p. 133) is immediately derived by effecting the process of variation.

If T is a homogeneous quadratic function of the generalized components of velocity, $T^{\frac{1}{2}}dt$ is a quadratic function of infinitesimal elements of the coordinates: therefore the first form when expressed (after Jacobi) as

$$\delta \int (E - W)^{\frac{1}{2}} (T^{\frac{1}{2}}dt) = 0$$

does not any longer involve the time. It thus determines the geometrical relations of the path of the system without reference to time; for a simple orbit it reduces to the earliest form investigated above.

In the modern discussions of the fundamental principles of dynamics, especially as regards their tentative adaptation to new regions of physical phenomena whose dynamical connexions are concealed, this principle of variation of the Action, which condenses the whole subject into a single formula independent of any particular system of coordinates, naturally occupies the most prominent place.

As a supplement to Chapter IX, these statements of the Principle of Action will now be established for a general dynamical system. This can be done most simply and powerfully by introducing the analytical method of Variations, invented by Lagrange as above mentioned.

The principle, as already deduced for the simplest case, relates to the forms of paths or orbits: if it is also to involve the manner in which the orbits are described the time must come in. The criterion of a free path was that $\delta \int v ds = 0$ with energy E_0 constant throughout the motion: it is the same as $\delta \int v^2 dt = 0$ under the same condition; or, writing T for the kinetic energy $\frac{1}{2}mv^2$, it is $\delta\int 2T dt = 0$ under the same restriction to constancy of the total energy.

Let us conduct the variation directly from this latter form, but now keeping the time unvaried,

$$\delta\int T dt = \delta\int_{t_1}^{t_2} \tfrac{1}{2}m\left\{\left(\frac{dx}{dt}\right)^2 + \left(\frac{dy}{dt}\right)^2 + \left(\frac{dz}{dt}\right)^2\right\} dt$$

$$= \int m\left(\frac{dx}{dt}\frac{d\delta x}{dt} + \frac{dy}{dt}\frac{d\delta y}{dt} + \frac{dz}{dt}\frac{d\delta z}{dt}\right) dt$$

in which d is the differential of x as the body moves along its orbit with changing time, but δx is the variation of the value of x as we pass from a point on the orbit to a corresponding point on the adjacent possible path that is compared with it. The introduction of different symbols d and δ to discriminate these two types of change is the essential feature of the Calculus of Variations: we have already used the fundamental relation $\delta dx = d\delta x$. Integrating now by parts, in order to get rid of variations of velocities which are not independent variations and so not arbitrary, we obtain

$$\delta\int T dt = \left| m\frac{dx}{dt}\delta x + m\frac{dy}{dt}\delta y + m\frac{dz}{dt}\delta z \right|_1^2$$

$$- \int\left(m\frac{d^2x}{dt^2}\delta x + m\frac{d^2y}{dt^2}\delta y + m\frac{d^2z}{dt^2}\delta z\right) dt;$$

in this the first term represents the difference of the values at the upper and lower limits of the integral, indicated by subscripts 2 and 1, which correspond to the final and initial positions of the body. The second term is equal to

$$- \int (X\delta x + Y\delta y + Z\delta z)\, dt,$$

where (X, Y, Z) is the effective force acting on the particle m, as determined by the acceleration which the particle acquires.

We can extend this equation at once to any system of particles in motion under both extraneous and mutual forces. If there are no forces exerted from outside the system, but only an internal potential energy expressed by a function W, then the work of the internal forces of the system tends to exhaust this energy, so that

$$\Sigma\,(X\delta x + Y\delta y + X\delta z) = -\,\delta W,$$

and this holds good whether the algebraic equations expressing the constraints contain t or not.

Thus if T now represents the total kinetic energy, and all the forces are internal, we can write, for variation from a free path to any adjacent path by frictionless constraint, and with times unvaried,

$$\delta \int (T - W)\, dt = \left| \Sigma m \frac{dx}{dt}\,\delta x + \Sigma m \frac{dy}{dt}\,\delta y + \Sigma m \frac{dz}{dt}\,\delta z \right|_1^2.$$

Strictly, this result has been obtained for a system of separate particles influencing each other by mutual forces. It is natural to expand it to any material system consisting of elements of mass subject to mutual forces, thus including the dynamics of elastic systems. The ultimate analysis of the element of mass is into molecules or atoms in a state of internal motion: that final extension would include the dynamical theory of heat.

We can now express all the coordinates x, y, z of the particles or elements of mass in terms of any sufficient number of independent quantities θ, ϕ, ψ, ... which determine the position and configuration of the system as restricted by its structure. Their number is that of the degrees of freedom of the system. The equations which express x, y, z in terms of them may involve t explicitly, for the equation of virtual work involves the displacements possible *at given time*; thus the new form of $T - W$ can contain t. Then we can assert that when t is not varied, and the time limits t_1 and t_2 are therefore constant,

$$\delta \int_{t_1}^{t_2} (T - W)\, dt = 0$$

when the frictionless variation is taken between fixed initial and final positions of the dynamical system.

This quantity $T - W$ is the Lagrangian function L defining by itself alone the dynamical character of the system: the function $-L$ or $W - T$ is thus the potential energy W as modified for kinetic applications, and has been appropriately named by Helmholtz the kinetic potential of the system. Thus the particular case of a system at rest is included: for

$$\delta \int W dt \quad \text{or} \quad \int \delta W dt \quad \text{is equal to} \quad \delta W \int dt$$

as W remains constant during the time: hence the equation of Action asserts in this case that

$$\delta W = 0,$$

which comprehends the laws of Statics in the form that the equilibrium is determined by making the potential energy stationary. For stability it must be minimum.

Again, as L is expressed as a function of the generalized coordinates $\theta, \phi, \ldots$ and their velocities,

$$\delta \int L dt = \int \left(\frac{\partial L}{\partial \theta} \delta\theta + \frac{\partial L}{\partial \dot\theta} \delta\dot\theta + \ldots \right) dt$$

where $\dot\theta$ represents $\frac{d\theta}{dt}$, and $\delta\dot\theta$ is equal to $\frac{d}{dt}\delta\theta$: thus integrating by parts as before

$$\delta \int_{t_1}^{t_2} L dt = \left| \frac{\partial L}{\partial \dot\theta} \delta\theta + \frac{\partial L}{\partial \dot\phi} \delta\phi + \ldots \right|_1^2$$

$$- \int \left\{ \left(\frac{d}{dt} \frac{\partial L}{\partial \dot\theta} - \frac{\partial L}{\partial \theta} \right) \delta\theta + (\ldots)\, \delta\phi + \ldots \right\} dt.$$

As the left side vanishes, when the terminal positions

are unvaried, for all values of the current variations $\delta\theta, \delta\phi, \ldots$, and these are all independent and arbitrary, the coordinate quantities $\theta, \phi, \ldots$ being just sufficient to determine the system, the coefficient of each of these variations must vanish separately in the integrand. Thus we obtain a set of equations of type

$$\frac{d}{dt}\frac{\partial L}{\partial \dot{\theta}} - \frac{\partial L}{\partial \theta} = 0$$

which are the Lagrangian equations of motion of any general dynamical system (20, p. 133 *supra*). If there are in addition extraneous forces in action on the system, the appropriate component force F_θ, defined as that part whose work $F_\theta \delta\theta$ is confined to change of the one coordinate θ, must be added on the right-hand side. These applied forces may vary with t in any manner: they can be merged in W by addition of terms $-F_\theta\theta - \ldots$ to it: their presence will prevent the energy of the system from remaining constant.

If we restrict this comparison of paths to variation from a free path of the system *to adjacent free paths*, we have

$$\delta \int_{t_1}^{t} L dt = \frac{\partial L}{\partial \dot{\theta}} \delta\theta + \frac{\partial L}{\partial \dot{\phi}} \delta\phi + \ldots$$

now as an *exact* equation, and so capable of further differentiation; and it provides the basis of the Hamiltonian theory of varying Action.

It will be convenient at this stage to remove the restriction that the time is not to be varied: to allow for this change we must substitute in the equation in place of $\delta\theta$ the expression $\delta\theta - \dot{\theta}\delta t$ which deducts from the total variation of θ that part of it which arises from the motion in the interval of varied time δt. We must also add $L\delta t$ in order to include in the time of transit the new interval of time δt added on at the end by the variation. Thus now

$$\delta \int_{t_1}^{t} L\delta t = L\delta t + \frac{\partial L}{\partial \dot{\theta}} (\delta\theta - \dot{\theta}\delta t) + \frac{\partial L}{\partial \dot{\phi}} (\delta\phi - \dot{\phi}\delta t) + \ldots.$$

Also $L = T - W$; and T being a homogeneous quadratic function,

$$\frac{\partial L}{\partial \dot{\theta}}\dot{\theta} + \frac{\partial L}{\partial \dot{\phi}}\dot{\phi} + \ldots = 2T;$$

hence $$\delta \int_{t_1}^{t} L dt = \frac{\partial L}{\partial \dot{\theta}}\delta\theta + \frac{\partial L}{\partial \dot{\phi}}\delta\phi + \ldots - E\delta t$$

where E is the final value of the total energy $T + W$.

When no extraneous forces are supposed to be in action E is constant at all times: thus

$$- E\delta t = t\delta E - \delta\,(Et) = t\delta E - \delta \int E dt.$$

Hence, transposing the last term, the alternative form arises,

$$\delta \int_{t_1}^{t} 2T dt = \frac{\partial T}{\partial \dot{\theta}}\delta\theta + \frac{\partial T}{\partial \dot{\phi}}\delta\phi + \ldots + t\delta E,$$

for variations throughout which the energy is conserved.

This is the generalization of the previous form $\delta \int mv ds = 0$ for a particle, except that now the time also is involved, and is determined as $\partial A/\partial E$, where A is the time integral of $2T$ as expressed in terms of initial and final configurations and the conserved energy.

This involves the analytical result that if Θ, Φ, ... are the momenta* corresponding to the coordinates $\theta, \phi, \ldots$, then there must exist a certain function A (of form however that is usually difficult to calculate) of $\theta, \phi, \ldots E$, such that in varying from the free path to adjacent free paths of the system,

$$\delta A = \Theta\delta\theta + \Phi\delta\phi + \ldots + t\delta E.$$

A more explicit and wider form, especially for optical applications, is immediately involved in this formula,

* The subscript notation of Chapter IX would here be inconvenient.

that there is a function $A\Big|_1^2$ of the initial and final configurations of the system and the energy, such that

$$\delta A\Big|_1^2 = \Theta_2\delta\theta_2 + \Phi_2\delta\phi_2 + \ldots \\ - \Theta_1\delta\theta_1 - \Phi_1\delta\phi_1 - \ldots + (t_2 - t_1)\,\delta E.$$

There also exists a function $P\Big|_1^2$ of the final and initial coordinates and the time, equal in value to $\int_{t_1}^{t_2}(T-W)\,dt$, such that

$$\delta P\Big|_1^2 = \Theta_2\delta\theta_2 + \Phi_2\delta\phi_2 + \ldots \\ - \Theta_1\delta\theta_1 - \Phi_1\delta\phi_1 - \ldots - E_2\delta t_2 + E_1\delta t_1,$$

on varying from any free orbit to adjacent free orbits; but now as there is no restriction to E remaining constant along an orbit, the forces may be in part extraneous forces whose work will impart new energy to the system.

The mere fact that such a function P or A exists involves a crowd of reciprocal differential relations connecting directly the initial and final configurations of the system or a group of systems, of type such as

$$\partial\Theta_1/\partial\phi_2 = -\,\partial\Phi_2/\partial\theta_1,$$

which are often the expression of important physical results. Moreover in the form of δP, and therefore in such resulting relations, the final set of coordinates may be different from the initial set.

The influence of disturbing agencies on any dynamical system, whose undisturbed path was known, is by these principles reduced to determining by approximation (from a differential equation which it satisfies) the slight change they produce in this single function P or A which expresses the system, a method perfect in idea but amenable to further simplifications in practice.

This beautiful theory of variation of the Action from any free path to the adjacent ones was fully elaborated by Hamilton in a single memoir in two parts (*Phil*

Trans., 1834 and 1835), and soon further expanded in analytical directions by Jacobi and other investigators. It brings a set of final positions of a dynamical system into direct relations with the corresponding initial positions, independently of any knowledge whatever of the details of the paths of transition. In connexion with the simplest case of orbits it has been characterized by Thomson and Tait as a theory of aim, connecting up, so to say, the deviations on a final target, arising from changes of aim at a firing point, with the corresponding quantities of the reversed motion. In geometrical optics, from which the original clue to the theory came, where the rays might be regarded as orbits of imagined Newtonian corpuscles of light, it involves the general relations of image to object that must hold for all types of instrument, as originally discovered by Huygens and by Cotes. Its scope now extends all through physical science.

In certain cases the number of coordinate variables required for the discussion of a dynamical problem can be diminished. Thus if the kinetic potential involves one or more coordinates only through their velocities, the corresponding equations of motion merely express the constancy throughout time of the momentum that is associated with each such coordinate: this holds for instance for the case of freely spinning flywheels attached to any system of machinery, and for all other cases in which configuration is not affected by the changing value of the coordinate. In all such cases the velocity can be eliminated, being replaced by its momentum which is a physical constant of the motion. The kinetic potential can thus be modified (Routh, Kelvin, Helmholtz) so as to involve one or more variables the less, but still to maintain the stationary property of its time-integral. It is now no longer a homogeneous quadratic, but involves terms containing the other velocities to the first degree, multiplied of course by these constant momenta as all the terms must be of the same dimensions. Every such kinetic potential belongs to a system possessing one or more *latent* unchanging (steady) motions; and a general theory of this important physical class of systems, and of the transformation of their energies, arises.

In fact if
$$L' = L - \Psi\dot{\psi} - \dots$$
where $\psi \dots$ are a group of coordinates and $\Psi \dots$ the related momenta, then
$$\delta L' = \left(\frac{\partial L}{\partial \dot{\psi}} - \Psi\right)\delta\dot{\psi} - \left(\dot{\psi} - \frac{\partial L}{\partial \Psi}\right)\delta\Psi + \frac{\partial L}{\partial \psi}\delta\psi + \dots + \delta_1 L$$
in which the first term vanishes identically, while $\delta_1 L$ is the variation of L with regard to the remaining variables. Hence if L do not involve the coordinates $\psi \dots$, so that $\Psi \dots$ are constant and are not made subject to variation, and $\dot{\psi} \dots$ are eliminated from L' by introduction of $\Psi, \dots$ then
$$\delta \int L'dt = \Big| \Theta\delta\theta + \Phi\delta\phi + \dots - E\delta t \Big|_1^2$$
depending only on the variations of the explicit coordinates at the limits, provided $\Psi \dots$ are kept unvaried, or the flywheels of the system are not tampered with.

Although the cyclic coordinates do not appear at all in L, yet it is only in terms of L' modified as here that we can avoid their asserting themselves in the domain of varying Action.

The ultimate aim of theoretical physical science is to reduce the laws of change in the physical world as far as possible to dynamical principles. It is not necessary to insist on the fundamental position which the kinetic potential and the stationary property of its time-integral assume in this connexion. Two dynamical systems whose kinetic potentials have the same algebraic form are thoroughly correlative as regards their phenomena, however different they may be in actuality. If any range of physical phenomena can be brought under such a stationary variational form, its dynamical nature is suggested: there still remains the problem to extricate the coordinates and velocities and momenta, and to render their relations familiar by comparison with analogous systems that are more amenable to inspection and so better known.

Note on Chapter IX, § 9.

It has appeared above, as Lagrange long ago emphasized, that the principle of Conservation of Energy can provide only one of the equations that are required to determine the motion of a dynamical system. It follows that the reasoning of this section (§ 9), which seems to deduce them all, must be insufficient. The argument there begins by supposing the system to move in any arbitrary way; that is, it assumes motions determined by the various possible types of frictionless constraint that are consistent with the constitution of the system. The equation (9) is then derived correctly from (7) and (8), as the variations δq are fully arbitrary. But the imposed constraints introduce new and unknown constraining forces which must be included in the *applied* forces F_r; and they would make the result, so far as there demonstrated, nugatory.

The equations (9) are however valid, though this deduction of them fails. As explained above, the Lagrangian equations (20) are derivable immediately from the Principle of Least Action, independently established as here: and then the equations (9) can be derived by reversing the argument.

The procedure of § 12 seems to lead to a noteworthy result. It asserts that if

$$F = T_p + T_{\dot{q}} - 2T_{p\dot{q}}$$

then the single relation

$$\delta F = 0$$

involves all the equations connecting coordinates, velocities and momenta in the system. This will remain true when the three sets of variables, regarded still as independent, are changed to new ones by any equations of transformation, so that this threefold classification into types becomes lost. Now there are cases in which the steady motion of a system, or an instantaneous phase of a varying mode of change, can be thoroughly explored experimentally, leading to the recognition say of $3n$ physical quantities of which only $2n$ can be independent; but it is not indicated by our knowledge how we are to deduce from them a scheme of n coordinates, n corresponding velocities, and n momenta. We have arrived at the result that in every such case a function F must exist, and is capable of construction, such that $\delta F = 0$ provides a set of $3n$ equations containing all the knowledge that is needed. The relations (treated after Maxwell) of a network of mutually influencing electric coils carrying currents would form an example.

In cognate manner we may assert another type of equation of Variation of Action

$$\delta \int (T_p - 2T_{p\dot{q}} + W)\, dt = 0$$

where $T_{p\dot{q}} = \frac{1}{2}\Sigma \dot{q} p$, containing n coordinates q, their n velocities $\dot{q}$ and their n momenta p. For this equation is equivalent to

$$\int \Sigma \left\{ \left(\frac{\partial T_p}{\partial p} - \dot{q} \right) \delta p - p \delta \dot{q} + \frac{\partial T_p}{\partial q} \delta q + \frac{\partial W}{\partial q} \delta q \right\} dt = 0$$

leading on integration by parts as usual to two sets of relations of the types

$$\frac{\partial T_p}{\partial p} = \dot{q}, \quad \frac{dp}{dt} = -\frac{\partial T_p}{\partial q} - \frac{\partial W}{\partial q}$$

if in it the momenta and coordinates are regarded as independent variables. As $\frac{\partial T_p}{\partial q} = -\frac{\partial T_{\dot{q}}}{\partial q}$ by (18), the second set are the Lagrangian dynamical equations (20).

Thus we have here a single function

$$\phi = T_p - 2T_{p\dot{q}} + W$$

involving coordinates and their velocities, linear in the latter, and an equal number of quantities p of the nature of momenta, the coordinates and momenta being thus independent variables, such that the relation

$$\delta \int \phi dt = 0$$

leads both to the identification of the relations in which the momenta stand to the coordinates and to the dynamical equations of motion of the system.

This result is virtually the same as equation 12 *a* in Hamilton, *Phil. Trans.*, 1835, p. 247. In Helmholtz's memoir on Least Action, *Crelle's Journal*, vol. 100 (1886), *Collected Papers*, vol. iii, p. 218 another function is introduced, apparently with less fitness, in which the velocities are regarded as independent of their coordinates but the momenta are the gradients of L with regard to the velocities. Cf. also *Proc. Lond. Math. Soc.*, 1884.

A main source of the great power of these dynamical relations of minimal or stationary value, as exploring agents in physical science, is that the results remain valid however the physical character of the functions involved may be disguised by transformation to new variables, given in terms of the more fundamental dynamical ones by any equations whatever. This function ϕ may thus be expressed in terms of $2n$ quantities which are in any way mixed functions of coordinates and momenta and their gradients with respect to time—remaining a linear function of the latter and subject to other limitation—and the equation $\delta \int \phi dt = 0$ will still subsist and will express all the dynamical relations of the physical system.

The existence of a variational relation of this type may be taken as the ultimate criterion that a partially explored physical system conforms to the general laws of dynamics; while from its nature the coordinate quantities, in terms of which the configuration and motion of the system happen to be expressed, shrink to subsidiary importance.

INDEX

The numbers refer to pages